Axel Angeli
Ulrich Streit
Robi Gonfalonieri

The SAP R/3® Guide to EDI and Interfaces

Other titles in Computing

Books already in print:

The Efficiency of Theorem Proving Strategies
by David A. Plaisted and Yunshan Zhu

Recovery in Parallel Database Systems
by Svein-Olaf Hvasshovd

Applied Pattern Recognition
by Dietrich W. R. Paulus and Joachim Hornegger

Efficient Software Development with DB2 for OS/390
by Jürgen Glag

Corporate Information with SAP®-EIS
by Bernd-Ulrich Kaiser

SAP® R/3® Interfacing using BAPIs
by Gerd Moser

Multiobjective Heuristic Search
by Pallab Dasgupta, P. P. Chakrabarti and S. C. DeSarkar

Intelligent Media Agents
by Hartmut Wittig

Scalable Search in Computer Chess
by Ernst A. Heinz

Joint Requirements Engineering
by Georg Herzwurm, Sixten Schockert and Werner Mellis

The SAP R/3® Guide to EDI and Interfaces
by Axel Angeli, Ulrich Streit and Robi Gonfalonieri

Vieweg

Axel Angeli
Ulrich Streit
Robi Gonfalonieri

The SAP R/3® Guide
to EDI and Interfaces

Cut your Implementation Cost
with IDocs®, ALE® and RFC®

Second, revised Edition

vieweg

Die Deutsche Bibliothek – CIP-Cataloguing-in-Publication-Data
A catalogue record for this publication is available from Die Deutsche Bibliothek
(http://www.ddb.de).

1st Edition 2000
2nd, revised Edition January 2001

Vieweg is a company in the specialist publishing group BertelsmannSpringer.

Printing and binding: Lengericher Handelsdruckerei, Lengerich
Printed on acid-free paper
Printed in Germany

ISBN 3-528-15729-1

For Doris, Paul, Mini, Maxi

Axel Angeli,

is born in 1961. He is a <u>Top Level SAP R/3 consultant</u> and <u>R/3 cross-application development coach</u>. He specializes in coaching of large multi-national, multi-language development teams and troubleshooting development projects.

His job description is also known as computer logistics, a delicate discipline that methodically wakes the synergetic effects in team to accelerate and mediate IT projects.

He is a learned Cybernetics scientist (also known as Artificial Intelligence) in the tradition of the Marvin Minsky [*The society of mind*] and Synergetics group of Herman Haken and Maria Krell. His competence in computer science is based on the works of Donald Knuth [*The Art of Computer Programming*], Niklas Wirth (the creator of the PASCAL language), the object oriented approach as described and developed during the XEROX PARC project (where the mouse and windows style GUIs have been invented in the early 1970ies) and Borland languages.

Before his life as SAP consultant, he made a living as a computer scientist for medical biometry and specialist for high precision industry robots. He concentrates now on big international projects. He speaks fluently several popular languages including German, English, French and Slavic.

✍ axela@logosworld.de

Robi Gonfalonieri,

born in 1964 is a senior ABAP IV developer and R/3 consultant for SD and MM. He is a learned economist turned ABAP IV developer. He specializes in international, multi-language projects both as developer and SD consultant. He speaks fluently several languages including German, French, English and Italian.

✍ robig@logosworld.de

Ulrich Streit,

born in 1974 is ABAP IV developer and interface specialist. He developed a serious of legacy system interfaces and interface monitors for several clients of the process industry.

✍ ulis@logosworld.de

logosworld.com

is a group of loosely related freelance R/3 consultants and consulting companies. Current members of the logosworld.com bond are the following fine companies:

- Logos! Informatik GmbH, Brühl, Germany: R/3 technical troubleshooting
- OSCo GmbH, Mannheim, Germany: SAP R/3 implementation partner
- UNILAN Corp., Texas: ORACLE implementation competence

For true international R/3 competence and enthusiastic consultants,
email us ✍ info@logosworld.de
or visit http://idocs.de

Danke, Thank You, Graçias, Tack så mycket, Merci, Bedankt, Grazie, Danjawad, Nandri, Se-Se

I due special thanks to a variety of people, clients, partners and friends. Their insistence in finding a solution and their way to ask the right questions made this book only possible.

I want especially honour *Francis Bettendorf*, who has been exactly that genre of knowledgeable and experienced IT professionals I had in mind, when writing this book. A man who understands an algorithm when he sees it and without being too proud to ask precise and well-prepared questions. He used to see me every day with the same phrase on the lips: "Every day one question." He heavily influenced my writing style, when I tried to write down the answers to his questions. He also often gave the pulse to write down the answers at all. At the age of 52, he joyfully left work the evening of Tuesday the 23rd March 1999 after I had another fruitful discussion with him. He entered immortality the following Wednesday morning. We will all keep his memory in our heart.

Thanks to *Detlef* and *Ingmar Streit* for doing the great cartoons.

Thanks also to Pete Kellogg of UNILAN Corp., Texas, Juergen Olbricht, Wolfgang Seehaus and his team of OSCo, Mannheim for continuously forming such perfect project teams. It is joy working with them.

Plans are fundamentally ineffective because the "circumstances of our actions are never fully anticipated and are continuously changing around us". Suchman does not deny the existence or use of plans but implies that deciding what to do next in the pursuit of some goal is a far more dynamic and context-dependent activity than the traditional notion of planning might suggest.

Wendy Suchman, Xerox PARC http://innovate.bt.com/showcase/wearables/

Who Would Read This Book?

This book was written for the experienced R/3 consultants, who wants to know more about interface programming and data migration. It is mainly a compilation of scripts and answers who arose during my daily work as an R/3 coach.

Quid – What is that book about?
The R/3 Guide is a *Frequently Given Answers* book. It is a collection of answers, I have given to questions regarding EDI over and over again, both from developers, consultants and client's technical staff. It is focussed on the technical aspect of SAP R/3 IDoc technology. It is not a tutorial, but a supplement to the R/3 documentation and training courses.

Quis – Who should read the book?
The R/3 Guide has been written with the experienced consultant or ABAP developer in mind. It does not expect any special knowledge about EDI, however, you should be familiar with ABAP IV and the R/3 repository.

Quo modo – how do you benefit from the book?
Well, this book is a "How to" book, or a "Know-how"-book. *The R/3 Guide* has its value as a compendium. It is not a novel to read at a stretch but a book, where you search the answer when you have a question.

Quo (Ubi) – Where would you use the book?
You would most likely use the book when being in a project involved in data interfaces, not necessarily a clean EDI project. IDocs are also helpful in data migration.

Quando – when should you read the book
The R/3 Guide is not a tutorial. You should be familiar with the general concept of IDocs and it is meant to be used after you have attended an R/3 course on IDocs, ALE or similar. Instead of attending the course you may alternatively read one of the R/3 IDoc tutorial on the market.

Cur – Why should you read the book
Because you always wanted to know the technical aspects of IDoc development, which you cannot find in any of the publicly accessible R/3 documentation.

http://idocs.de http://logosworld.com

The A-Team of R/3 development

This book is the outcome of many years of work with and for SAP R/3. It could not have happened without the help of my colleagues and friends from logosworld.com. logosworld.com is today an informal association of freelance developers and experienced R/3 advisors in the technical field.

We may feel as outsiders by claiming that we are not only working for money but also with a big amount of idealism. We are mainly interest in demanding, challenging and pioneering tasks. Solving a cunning riddle is much more interesting for us than setting up the same system for the hundredth time, just for the sake of the good money.

We have seen many projects, where some consulting firm appeared with twenty consultants, non of them experienced enough to really push a project. The mere quantity should help fulfilling the time lines.

The contrary was true, of course. Two skilled consultants with three or four intelligent helpers at hand would have set up the system in half the time and with a better result. How would that be possible? R/3 is still a field, where the time to solve a problem can span enormously. We have many examples where we could have solved a problem in even ten percent of the time than the ordinary consultant team did.

Where do we save the time? First, we do not have to experiment, because we have many scenarios in our mind and because we know the coding, we can already predict how the system behaves and what is possible. Secondly, we hate conferences. Most of what needs to be said, can be said during working. A conference does not solve a problem at all.

A cute manager told me once the story: if you have a problem and do not know a solution, then you do not really have a problem, because you are only whining.

We are not afraid of making a mistake. We will clean up the mess. It is better to ask for forgiveness, then for permission.

We would like to be your partner, if you ask for quick, stable and efficient technical solutions and if you think that your project burst all limits or is in jeopardy.

We cannot save the world for you, but we can save your project, when others fail, because we know R/3 from the inside. If there is something not right with your system, we are like a mechanic who opens the lid and looks at the engine, while others may still walk in circles around the installation, guessing and horoscoping. That is why we feel like the A-Team for R/3 development:

See you at http://logosworld.com .

NB: All educational information on logosworld.com is still free and will remain free.

Directory of Programs

http://idocs.de http://logosworld.com

Table of Contents

Summary

http://idocs.de http://logosworld.com

http://idocs.de http://logosworld.com

EDI Converter

Appendix

Index

Table of Illustrations

http://idocs.de http://logosworld.com

Where Has the Money Gone?

EDI projects can soon become very expensive. However, when analysing the reasons for high costs, one finds quickly that it is not the technical implementation of the EDI project that lets explode the total costs.

Summary

- Most of the implementation time and costs get lost in agreeing common standards and establishing formalism between the sender and the receiver
- A successful EDI project requires the developers on both ends sitting together face to face
- Sticking to a phantom "SAP standard" for IDocs, which does not actually exist in R/3, lets the costs of the project soar

Just make a plan,	*Mach nur einen Plan,*
And let your spirit hail.	*Sei ein großes Licht,*
Then you make another plan,	*Dann mach noch*
	einen zweiten Plan
And both will fail.	*Gehen tun sie beide nicht.*

Bertold Brecht and Kurt Weill, Three Penny Opera

1.1 Communication

More than 80% of the time of an EDI project is lost in waiting for answers, trying to understand proposals and retrieving data nobody actually needs.

A common language

EDI means to exchange information between a sender and a receiver. Both communication partners need to speak the same language to understand each other.

The language for EDI are the file formats and description languages used in the EDI data files. In the simple case of exchanging plain data files, the partners need to agree on a common file format.

Finding the common agreement, that is it, where most of the money gets lost. See a common scenario:

The receiving party defines a file structure in which it likes to receive the data. This is usually an image of the data structure of the receiving computer installation.

This is a good approach for the beginning, because you have to start somewhere. But now the disaster takes course.

The proposal is sent to the other end via email. The developer of the sender system takes a look on it and remains quiet. Then he starts programming and tries to squeeze his own data into the structure.

Waiting for a response

If it becomes too tedious, a first humble approach takes place to convince the other party to change the initial file format. Again it is sent via email and the answer comes some days later. Dead time, but the consultant is paid.

Badly described meaning of a field

It can be even worse: one party proposes a format and the other party does not understand the meaning of some fields.

Echoing

Another field cannot be filled, because the sender does not have the information. Looking closer you find out, that the information originates from the receiving partner anyway. The programmer who proposed the format wanted it filled just for his personal ease. This is known as **Echoing** and it is always a nice to have feature.

Using the same term for different objects

A real disaster happens if both parties use the same expression for different items. A classy case is the term "**delivery**": many legacy systems call a delivery what is known as an SD transport in R/3.

There are many other situation where always one thing happens: time is spoiled. And time is money.

Face to face

The solution is more than easy: bring the people together. Developers of both parties need to sit together, physically face to face. If they can see what the other person does, they understand each other.

1.2   Psychology of Communication

Bringing developers together accelerates every project. Especially when both parties are so much dependent on each other as in an EDI project, the partners need to communicate without pause.

There is a psychological aspect in the communication process, if the parties on both ends do not know each other or reduce communication with each other to the absolute minimum,

Sporadic communication leads to latent aggression on both sides, while spending time together builds up mutual tolerance. Communicating directly and regularly, rises pretty certainly the mutual respect. Once the parties accept the competence of each other they accept the other's requirements more easily.

Send them over the ocean.

Why, will you say, what if people sit on two ends of the world, one in America the other in Europe? The answer is strict and clear: get them a business class flight and send them over the ocean.

Travel cost will be refunded by the saved time

The time you will save when the people sit together will even up a multitude of the travel costs. So do not think twice.

Sitting together also rises the comprehension of the total system. An EDI communication forms a logical entity. But if your left hand does not know what your right hand does, you will never handle things firm and secure.

See the business on both ends

Another effect is thus a mutual learning. It means to learn how the business is executed on both sides. Seeing the commons and the differences allows flexibility. And it allows to make correct decisions without needing to ask the communication partner.

http://idocs.de http://logosworld.de

1.3 Phantom SAP Standards and a Calculation

SAP R/3 delivers a serious of predefined EDI programs. Many project administrators see them as standards which should not be manipulated or modified. The truth is, that these IDoc processing functions are recommendations and example routines, which can be replaced be own routines in customizing.

Predefined not standard

SAP R/3 is delivered with a serious of predefined IDoc types and corresponding handler function modules.

Some of the handler programs had been designed with user-exits where a developer could implement some data post-processing or add additional information to an IDoc.

You must always see those programs as examples for IDoc handling. If the programs do already what you want, it is just fine. But you should never stick too long to those programs, if you need different data to be sent.

R/3 IDocs were primarily designed for the automotive industry

The R/3 standard IDoc programs had been designed with the German association of automobile manufacturers (VDA) in mind. The VDA is a committee which defines EDI standards for their members, e.g. Volkswagen, BMW, Daimler-Benz-Chrysler. Not every car manufacturer, e.g. FORD uses these recommendations. Other industries define their own standards which are not present in R/3.

If there already exists a file exchange format for your company or your industry, you may want to use this one. This means to type in the file format, writing the program that fills the structure and customize the new IDoc and message types.

A simple calculation:

Calculation

Discussing the solutions	5 days
Typing in the file formats	1/2 day
Writing the program to fill the segments	1 days
Adjust the customizing	1/2 day
Testing and correcting everything	3 days
Travel time	2 days
Total	12 days

This is not an optimistic calculation. You will mind that eight out of the twelve days are accounting for non IT related tasks like discussing solutions, educating each other and testing.

If a project takes longer than that, it always adds to the account of discussion and adapting solutions, because things have changed or turned out to be different as initially planned.

1.4 Strategy

Do not loose your time in plans. Have prototypes developed and take them as a basis.

You cannot predict all eventualities

Do not stick to the illusion, that a proper design in the beginning will lead to a good result. It is the age old error in trusting the theorem of Laplace:

Laplace

> *"Tell me all the facts of the world about the presence and I will predict the future for you."*

Heisenberg and uncertainty

Let aside the fact, that modern physics since Heisenberg and his uncertainty theorem has proven, that even knowing everything about now, does not allow to predict the future deterministically.

You do not know the premises before

If you want to know all the eventualities of a project, you have to be gone through similar projects. It is only your experience that allows you to make a good plan. However, you usually do a project only once, unless you are a consultant.

The question is: If you have never been through an EDI project, how will you obtain the necessary experience?

Prototypes

The answer is: make a prototype, a little project. Do not loose your time in writing plans and detailed development requests. Rather start writing a tiny prototype. Introduce this prototype and maintain your solution. Listen to the arguments and improve the prototype steadily.

This is how you learn.

This is how you succeed.

1.5 Who Is On Duty?

Writing interface programs is much like translating languages. The same rule apply.

Writing interface programs is like translating a language. You have information distributed by one system and you have to translate this information into a format that the other system understands.

A translation should always be done by a native speaker of the target language. This applies to interface programs as well.

If data needs to be converted, do this always in the target system. If in doubt let the source system send everything it can. If the target does not need the information it can ignore it.

http://idocs.de http://logosworld.de

1.6   Marcus T. Cicero

Some may have learned it in school: the basic rules of rhetoric according to Cicero. You will know the answers, when your program is at its end. Why don't you ask the questions in the beginning? Ask the right question, then you will know.

When starting a new task, you have always to answer the magic "Q" s of rhetoric. It is a systematic way to get the answer you need to know anyway.

Quid – What   What is the subject you are dealing with? Make clear the context you are in and that all parties talk about the same.

Quis – Who   Who is involved in the business? Get the names and make sure, that they know each other before the project enters the hot phase.

Quo modo – how   How do you want to achieve your goal? Be sure all participants choose the same methods. And how do you name the things? Agree on a common terminology!

Quo (Ubi) – where   Where do things take place? Decide for a common place to work. Decide the platform, where elements of the programs should run.

Quando - when   When do you expect a result? Define milestones and discuss the why when the milestones were missed. You should always check why your initial estimate was wrong, also if you are faster than planned.

Cur – Why   Why do you want to install a certain solution? Isn't there a better alternative?

2

What Are SAP R/3 IDocs?

IDocs are SAP's file format to exchange data with a foreign system. This chapter is intended as an introduction to the concept.

Summary

- IDocs are an ASCII file format to exchange data between computers; the format is chosen arbitrarily

- IDocs are similar to segmented files; they are <u>not</u> a description language like ANSI X.12, EDIFACT or XML

- The IDoc contents are processed by function modules, which can be assigned in customizing

http://idocs.de http://logosworld.de

2.1   What Are IDocs?

IDocs are structured ASCII files (or a virtual equivalent). They are the file format used by SAP R/3 to exchange data with foreign systems.

IDocs Are SAP's implementation of structured text files

IDocs are simple ASCII data streams. When they are stored to a disk file, the IDocs are simple flat files with lines of text, where the lines are structured into data fields. The typical structured file has records, where each record starts with a leading string, which identifies the record type. Their specification is stored in the data dictionary.

Electronic Interchange Document, Electronic Intermediate Document

IDocs is the acronym for Intermediate Document. This indicates a set of (electronic) information which build a logical entity. An IDoc is e.g. all the data of a single customer in your customer master data file. Or the IDoc is all the data of a single invoice.

Data Is transmitted in ASCII format, i.e. human readable form

IDoc data is usually exchanged between systems and partners who are completely independent. Therefore the data should be transmitted in a format, that can easily be corrected by the humans who operate the computers. It is therefore mandatory to post the data in a human readable form.

Nowadays, this means that data is coded in ASCII format, including number, which are sent as string of figures 0 to 9. Such data can easily be read with any text editor on any computer, be it a PC, Macintosh, UNIX System, S/390 or any internet browser.

IDocs exchange messages

The information which is exchanged by IDocs is called a message and the IDoc is the physical representation of such a message. The name "messages" for the information sent via IDocs is used in the same ways as other EDI standards do.

IDocs are used like classical interface files

Everybody who ever dealt with interface programming, will find IDocs very much like the hierarchical data files used in traditional data exchange.

International standards like the ODETTE or VDA formats are designed in the same way as IDocs are.

XML, ANSI X:12 or EDIFACT use a description language

Other EDI standards like XML, ANSI X.12 or EDIFACT/UN are based on a data description language. They differ principally from the IDocs concept, because they use a programming language syntax (e.g. like Postscript or HTML) to embed the data.

2.2 Exploring a Typical Scenario

The IDoc process is a straight forward communication scenario. A communication is requested, then data is retrieved, wrapped and sent to the destination in a predefined format and envelope.

The illustration above displays a sketch for a typical IDoc communication scenario. The steps are just the same as with every communication scenario. There is a requesting application, a request handler and a target.

The sketch shows the communication outbound R/3. Data is leaving the R/3 system.

R/3 application creates data

An R/3 application creates data and updates the database appropriately. An application can be a transaction, a stand-alone ABAP Report or any tool that can update a database within R/3.

IDoc engine picks up the request

If the application thinks that data needs to be distributed to a foreign system, it triggers the IDoc mechanism, usually by leaving a descriptive message record in the message table NAST.

The application then either calls directly the IDoc engine or a collector job eventually picks up all due IDoc messages and determines what to do with them.

IDoc engine determines a handler function from customizing

If the engine believes that data is fine to be sent to a partner system, then it determines the function module which can collect and wrap the required IDoc data into an IDoc.

In IDoc customizing, you specify the name of the function module to use. This can either be one which is predefined by R/3 standard or a user-written one.

IDoc is backup up in R/3 and sent out

When the IDoc is created it is stored in an R/3 table and from there it is sent to the foreign system.

Conversion to standards is done by external program

If the foreign system requires a special conversion, e.g. to XML, EDIFACT or X.12 then this job needs to be done by an external converter, like the Seeburger ELKE™ system. These converters are not part of R/3.

If you have to decide for a converter solution, we strongly recommend to use a plain PC based solution. Conversion requires usually a lot of fine tuning which stands and falls with the quality of the provided tools.

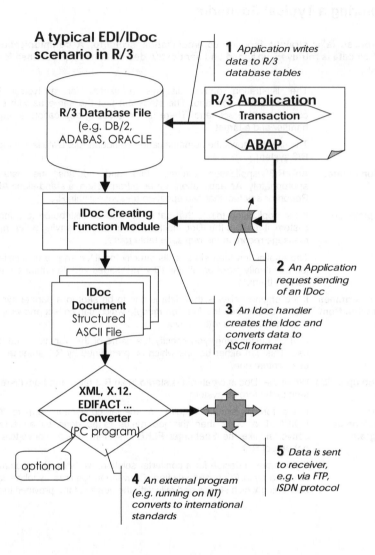

A typical EDI/IDoc scenario in R/3

R/3 Database File
(e.g. DB/2, ADABAS, ORACLE

R/3 Application
Transaction
ABAP

1 *Application writes data to R/3 database tables*

IDoc Creating Function Module

2 *An Application request sending of an IDoc*

IDoc Document
Structured
ASCII File

3 *An Idoc handler creates the Idoc and converts data to ASCII format*

XML, X.12. EDIFACT ... Converter (PC program)

optional

4 *An external program (e.g. running on NT) converts to international standards*

5 *Data is sent to receiver, e.g. via FTP, ISDN protocol*

Illustration 1: **A typical EDI scenario from the viewpoint of R/3**

| 3 |

Get a Feeling for IDocs

IDocs are relatively simple to understand. But, like most simple things they are difficult to explain. In this chapter we want to look on some IDoc and describe its elements, so that you can get a feeling for them.

Summary

- The first record in an IDoc is a control record describing the content of the data

- All but the first record are data records with the same formal record structure

- Every record is tagged with the segment type and followed by the segment data

- The interpretation of the segment is done by the IDoc application

- Both sent and received IDocs are logged in R/3 tables for further reference and archiving purposes

http://idocs.de http://logosworld.de

3.1 Get A Feeling For IDocs

For the beginning we want to give you a feeling of what IDocs are and how they may look like, when you receive it as a plain text file.

IDocs are plain ASCII files (resp. a virtual equivalent)

IDocs are basically a small number of records in ASCII format, building a logical entity. It makes sense to see an IDoc as a plain and simple ASCII text file, even if it might be transported via other means.

Control record plus many data records = 1 IDoc

Any IDoc consists of two sections
- The control record

which is always the first line of the file and provides the administrative information.
The rest of the file is made up by
- the data record

which contain the application dependent data, in our example below the material master data.

For an example, we will discuss the exchange of the material master IDoc MATMAS in the paragraphs below.

IDocs are defined in WE31

The definition of the IDoc structure MATMAS01 is deposited in the data dictionary and can be viewed with WE30 .

IDOC Number	Sender	Receiver	Port	Message Type	IDoc Type
0000123456	R3PARIS	R3MUENCHEN	FILE	ORDERS	ORDERS01

Illustration 2:   **Simplified Example of an IDoc control record for sales orders**

SegmentType	Sold-To	Ship-To	Value	Deldate	User
ORDERHEADER	1088	1089	12500,50	24121998	Micky Maus

Illustration 3:   **Simplified Example of an IDoc data record for sales orders**

```
EDI_DC40  0430000000001234540B 3012 MATMAS03      MATMAS    DEVCLNT100 PROCLNT100
E2MARAM001 0430000000001234500000100000002005TESTMAT1          19980303ANGELI          19981027SAPOSS
E2MAKTM001 0430000000001234500000300001030005EEnglish Name for TEST Material 1                    EN
E2MAKTM001 0430000000001234500000400001030005FFrench Name for TEST Material 1                     FR
E2MARCM001 0430000000001234500000500001030005010DEAVB        901          PD9010 0    0.00 EXX  0.000
E2MARDM001 0430000000001234500000600000050400051000D                   0.000       0.000
E2MARDM001 0430000000001234500000700000050400051200D                   0.000       0.000
E2MARMM    0430000000001234500000900001030005KGM1        1              0.000       0.000
```

Illustration 1:   Part of the content of an IDoc file for IDoc type MATMAS01

```
00000000012345 DEVCLNT100 PROCLNT100 19991103 210102
E1MARAM        005 TESTMAT1     19980303 ANGELI      19981027SAPOSS
  E1MAKTM      005 D German  Name for TEST Material 1                    DE
  E1MAKTM      005 E English Name for TEST Material 1                    EN
  E1MAKTM      005 F French  Name for TEST Material 1                    FR
  E1MARCM      005 0100 DEAVB                         901
  E1MARCM      005 0150 DEAVB                         901
    E1MARDM         005 1000 D              0.000     0.000
    E1MARDM         005 1200 D              0.000     0.000
  E1MARMM      005 KGM 1  1
  E1MARMM      005 PCE 1  1
```

Illustration 2:   The same IDoc in a formatted representation

3.2 The IDoc Control Record

The very first record of an IDoc package is always a control record. The structure of this control record is the DDic structure EDIDC and describes the contents of the data contained in the package.

Control record serves as cover slip for the transport

The control record carries all the administrative information of the IDoc, such as its origin and its destination and a categorical description of the contents and context of the attached IDoc data. This is very much like the envelope or cover sheet that would accompany any paper document sent via postal mail.

Control record is used by the receiver to determine the processing algorithm

For R/3 inbound processing, the control record is used by the standard IDoc processing mechanism, to determine the method how to process the IDoc. This method is usually a function module, but may be a business object as well. The processing method can be fully customized.

Control record not necessary to process the IDoc Data

Once the IDoc data is handed over to a processing function module, you will no longer need the control record information. The function modules are aware of the individual structure of the IDoc type and the meaning of the data. In other words: for every context and syntax of an IDoc, you would write an individual function module or business object (note: a business object is also a function module in R/3) to deal with.

Control Record structure is defined as EDIDC in DDic

The control record has a fixed pre-defined structure, which is defined in the data dictionary as EDIDC and can viewed with SE11 in the R/3 data dictionary. The header of our example will tell us, that the IDoc has been received from a sender with the name PROCLNT100 and sent to the system with the name DEVCLNT100 . It further tells us that the IDoc is to be interpreted according to the IDoc definition called MATMAS01 .

```
MATMAS01  ...  DEVCLNT100 PROCLNT100  ...
```

Illustration 4: **Schematic example of an IDoc control record**

Sender

The sender's identification PROCLNT100 tells the receiver who sent the IDoc. This serves the purpose of filtering unwanted data and gives also the opportunity to process IDocs differently with respect to the sender.

Receiver

The receiver's identification DEVCLNT100 should be included in the IDoc header to make sure, that the data has reached the intended recipient.

IDoc Type

The name of the IDoc type MATMAS01 is the key information for the IDoc processor. It is used to interpret the data in the IDoc records, which otherwise would be nothing more than a sequence of meaningless characters.

3.3 The IDoc Data

All records in the IDoc, which come after the control record are the IDoc data. They are all structured alike, with a segment information part and a data part which is 1000 characters in length, filling the rest of the line.

All IDoc data record have a segment info part and 1000 characters for data

All records of an IDoc are structured the same way, regardless of their actual content. They are records with a fixed length segment info part to the left, which is followed by the segment data, which is always 1000 characters long.

IDoc type definition can be edited with `WE30`

We will have a look on an IDoc of type `MATMAS01` . The IDoc type `MATMAS01` is used for transferring material master data via ALE. You can view the definition of any IDoc data structure directly within R/3 with transaction `WE30`.

Segment Info	Segment Data-→	
...E1MARAM	....00000001234567...	Material base segment
...E1MARCM	....PL01...	Plant Segment
...E1MARDM	....SL01	Storage location data
...E1MARDM	....SL02	Another storage location
...E1MARCM	....PL02	Another plant

Illustration 5: Example of an IDoc with one segment per line, an info tag to the left of each segment and the IDoc data to the right

Data and segment info are stored in `EDID4`

Regardless of the used IDoc type all IDocs are stored in the same database tables `EDID4` for release 4.x and `EDID3` for release 2.x and 3.x. Both release formats are slightly different with respect to the lengths of some fields. Please read the chapter on port types for details.

Depending on the R/3 release the IDoc data records are formatted either according the DDic structure `EDID3` or `EDID3`. The difference between the two structures reflects mainly the changes in the R/3 repository, which allow longer names starting from release 4.x.

http://idocs.de http://logosworld.de

3.4　Interpreting An IDoc Segment Info

All IDoc data records are exchanged in a fixed format, regardless of the segment type. The segment's true structure is stored in R/3's repository as a DDic structure of the same name.

R/3 is only interested in the segment name

The segment info tells the IDoc processor how the current segment data is structure and should be interpreted. The information, which is usually of only interest is the name of the segment `EDID4-SEGNAM`.

Segment name tells the data structure

The segment name corresponds to a data dictionary structure with the same name, which has been created automatically when defining the IDoc segment definition with transaction `WE31`.

Remaining information is only for foreign systems

For most applications, the remaining information in the segment info can be ignored as being redundant. Some older, non-SAP-compliant partners may require it. E.g. the IDoc segment info will also store the unique segment number for systems, which require numeric segment identification.

To have the segment made up for processing in an ABAP, it is usually wise to move the segment data into a structure, which matches the segment definition.

For a segment of type `e1maram` the following coding is commonly used:

Data in EDID4-SDATA

```
TABLES: e1maram.
     . . .
MOVE edidd-sdata TO e1maram.
```

Then you can access the fields of the IDoc segment `EDIDD-SDATA` as fields of the structure `e1maram`.

Data in EDID4-SDATA

```
WRITE: e1maram-matnr.
```

Sample coding

The following coding sample, shows how you may read a `MATMAS` IDoc and extract the data for the `MARA` and `MARC` segments to some internal variables and tables.

```
DATA: xmara LIKE e1maram.
DATA: tmarc AS STANDARD TABLE OF e1marcm
              WITH HEADER LINE.
LOOP AT edidd.
   CASE edidd-segnam.
      WHEN 'E1MARAM'.
           MOVE edidd-sdata TO xmara.
      WHEN 'E1MARCM'.
         MOVE edidd-sdata TO tmarc.
           APPEND tmarc.
   ENDCASE.
ENDLOOP.
now do something with xmara and tmarc.
```

3.5   IDoc Base - Database Tables Used to Store IDocs

When R/3 processes an IDoc via the standard inbound or outbound mechanism, the IDoc is stored in the tables. The control record goes to table `EDIDC` and the data goes to table `EDID4`.

All inbound and outbound Docs are stored in EDID4

All IDoc, whether sent or received are stored in the table `EDID4`. The corresponding control file header go into `EDIDC`.

There are standard programs who read and write the data to and from the IDoc base. These programs and transaction are heavily dependent on the customizing, where rules are defined which tell how the IDocs are to be processed.

Avoid reinventing the wheel

Of course, as IDocs are nothing than structured ASCII data, you could always process them directly with an ABAP. This is certainly the quick and dirty solution, bypassing all the internal check and processing mechanism. We will not reinvent the wheel here.

Customizing is done from the central menu WEDI

To do this customizing setting, check with transaction `WEDI` and see the points, dealing with ports, partner profiles, and all under IDoc development.

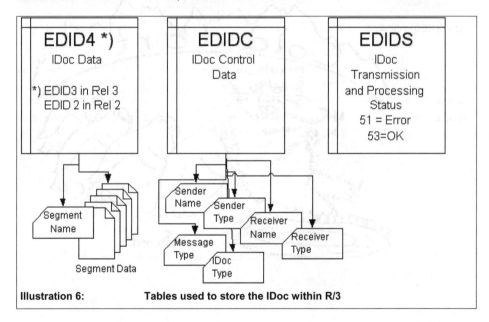

Illustration 6:   **Tables used to store the IDoc within R/3**

4

Exercise: Setting Up IDocs

The best way of learning is doing it. This chapter tells you how to set up your R/3 system that it can send IDocs to itself. When sending IDocs to your own system you can test the procedures without the need for a second client or installation.

Summary

- Define a new internal RFC destination INTERNAL
- Explore both the transactions WEDI and SALE and adjust the settings as necessary
- Use transaction BALE to generate an arbitrary IDoc

4.1 Quickly Setting up an Example

If you have a naked system, you cannot send IDocs immediately. This chapter will guide you through the minimum steps to see how the IDoc engine works.

You can access most of the transactions used in the example below in the menu WEDI and SALE.

Check EDID4 with SE16

We will assume, that we want to send material master data from the current system to a remote system. To simulate this scenario we do not need to have a second system. With a little trick, we can set up the system to send an IDoc back to sending client.

We will set up the system to use an RFC call to itself. Therefore we need to define an RFC remote destination, which points back to our own client. There is a virtual RFC destination called *NONE* which always refers to the calling client.

Declare the RFC destination to receive the IDoc

RFC destinations are installed with the transaction SM59. Create a new R/3 destination of type "L" (Logical destination) with the name *INTERNAL* and the destination *NONE*.
Note: Do not use RFC type internal. Although you could create them manually, they are reserved for being automatically generated. However, there is the internal connection "NONE" or "BACK" which would do the same job as the destination we are creating now.

Define a data port for *INTERNAL*

The next step is defining a data port, which is referenced by the IDoc sending mechanism to send the IDoc through. Declaring the port is done by transaction WE21.

Declare a new ALE model with SALE .

We will now declare an ALE connection from our client to the partner *INTERNAL*. ALE uses IDocs to send data to a remote system. There is a convenient transaction to send material master as IDocs via the ALE.

Declare *MATMAS01* as a valid ALE object to be sent to *INTERNAL*

The set up is done in transaction SALE. You first create a new ALE model, to avoid interfering with eventual existing definitions. Then you simply add the IDoc message *MATMAS* as a valid path from your client to *INTERNAL*.

Send the IDoc with transaction BALE.

In order to send the IDoc, you call the transaction BALE and choose the distribution of material master data (BD10). Choose a material, enter *INTERNAL* as receiver and go.

Display IDocs with WE05

To see, which IDocs have been sent, you can use the transaction WE05. If you did everything as described above, you will find the IDocs with an error status of *29*, meaning that there is no valid partner profile. This is true, because we have not defined one yet.

4.2 Example: The IDoc Type *MATMAS01*

To sharpen your understanding, we will show you an example of an IDoc of type *MATMAS01*, which contains material master data.

IDoc structure can be seen with WE30

Note: You can check with transaction WE05 , if there are already any IDocs in your system.

You can call transaction WE30 to display the structure of the IDoc type of the found IDoc.

Here is the display of an IDoc of type *MATMAS01*.

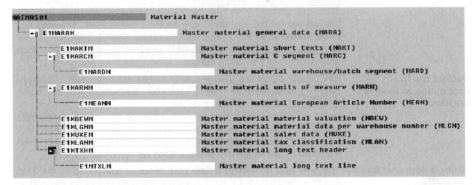

Structure of the MATMAS01 IDoc type

Content of IDoc file

MATMAS01 mirrors widely the structure of R/3's material master entity.

If this IDoc would have been written to a file, the file content would have looked similar to that:

```
...MATMAS01 DEVCLNT100 INTERNAL...
...E1MARAM ...and here the data
...E1MARCM ...and here the data
...E1MARDM ...and here the data
```

4.3   Example: The IDoc Type *ORDERS01*

To allow an interference, here is a sample of IDoc type *ORDERS01* which is used for purchase ordersand sales orders.

ORDERS01 is used for purchasing and sales order data

Purchasing and sales share naturally the same IDoc type, because what is a purchase order on sender side will become a sales order on the receiver side.

Other than *MATMAS01*, the IDoc type *ORDERS01* does not reflect the structure of the underlying RDB entity, neither the one of SD (VA01) nor the one of MM (ME21). The structure is rather derived from the EDI standards used in the automobile industry. Unfortunately, this does not make it easier to read.

Note: With transaction WE05 you can check, if there are already any IDocs in your system.

IDoc structure can be seen with WE30

You can call transaction WE30 to display the structure of the IDoc type of the found IDoc

Content of IDoc file

If this IDoc would have been written to a file, the file content would have looked similar to that:

```
...ORDERS01 DEVCLNT100 INTERNAL...
...E1EDKA1 ....and here the data
...E1EDKA2 ....and here the data
...E1EDP19 ....and here the data
```

ORDERS01	Purchasing/Sales
E1EDK01	IDoc: Document header general data
E1EDK14	IDoc: Doc.header organizational data
E1EDK03	IDoc: Document header date segment
E1EDK04	IDoc: Document header taxes
E1EDK05	IDoc: Document header conditions
E1EDKA1	IDoc: Doc.header partner information
E1EDK02	IDoc: Document header reference data
E1EDK17	IDoc: Doc.header terms of delivery
E1EDK18	IDoc: Doc.header terms of payment
E1EDKT1	IDoc: Doc.header text ID
E1EDKT2	IDoc: Doc.header texts
E1EDP01	IDoc: Doc.item general data
E1EDP02	IDoc: Doc.item reference data
E1EDP03	IDoc: Doc.item date segment
E1EDP04	IDoc: Doc.item taxes
E1EDP05	IDoc: Doc.item conditions
E1EDP20	IDoc schedule lines
E1EDPA1	IDoc: Doc.item partner information
E1EDP19	IDoc: Doc.item object identification
E1EDP17	IDoc: Doc.item terms of delivery
E1EDP18	IDoc: Doc.item terms of payment
E1EDPT1	IDoc: Doc.item text ID
E1EDPT2	IDoc: Doc.item texts

Illustration 7:   **Structure of the ORDERS01 IDoc type**

21

Sample Processing Routines

This chapter demonstrates on an example how an IDoc is prepared in R/3 for outbound and how a receiving R/3 system processes the IDoc.

Keep
It
Simple and
Smart

5.1 Sample Processing Routines

Creating and processing IDocs are a widely mechanical task, as it is true for all interface programming. We will show a short example that packs SAP R/3 SAPscript standard text elements into IDocs and stores them back.

Outbound function
Outbound IDocs from R/3 are usually created by a function module. This function module is called by the IDoc engine. A sophisticated customizing lets define the conditions and parameters to find the correct function module.

The interface parameters of the processing function need to be compatible with a well defined standard, because the function module will be called from within another program.

Inbound function
IDoc inbound functions are function modules with a standard interface, which will interpret the received IDoc data and prepare them for processing.

The received IDoc data is processed record by record and interpreted according the segment information provided with each record. The prepared data can then be processed by an application, a function module or a self-written program.

The example programs in the following chapters will show you how texts from the text pool can be converted into an IDoc and processed by an inbound routine to be stored into another system.

The following will give you the basics to understand the example:

Text from READ_TEXT
SAP R/3 allows the creation of text elements, e.g. with transaction SO10. Each standard text element has a control record which is stored in table STXH. The text lines themselves are stored in a special cluster table. To retrieve the text from the cluster, you will use the standard function module `function READ_TEXT` . We will read such a text and pack it into an IDoc. That is what the following simple function module does.

If there is no convenient routine to process data, the easiest way to hand over the data to an application is to record a transaction with transaction SHDB and create a simple processing function module from that recording.

Outbound is triggered by the application
Outbound routines are called by the triggering application, e.g. the RSNAST00 program.

Inbound is triggered by an external event
Inbound processing is triggered by the central IDoc inbound handler, which is usually the function module `IDOC_INPUT` . This function is usually activated by the gatekeeper, who receives the IDoc.

23

5.2 Sample Outbound Routines

The most difficult work when creating outbound IDocs is the retrieval of the application data which needs sending. Once the data is well retrieved, the data needs to be converted to IDoc format, only.

```
FUNCTION
*"----------------------------------------------------------------
*"*"Lokale Schnittstelle:
*"      IMPORTING
*"            VALUE(I_TDOBJECT) LIKE  THEAD-TDOBJECT DEFAULT 'TEXT'
*"            VALUE(I_TDID) LIKE  THEAD-TDID DEFAULT 'ST'
*"            VALUE(I_TDNAME) LIKE  THEAD-TDNAME
*"            VALUE(I_TDSPRAS) LIKE  THEAD-TDSPRAS DEFAULT SY-LANGU
*"      EXPORTING
*"            VALUE(E_THEAD) LIKE  THEAD STRUCTURE  THEAD
*"      TABLES
*"            IDOC_DATA STRUCTURE  EDIDD OPTIONAL
*"            IDOC_CONTRL STRUCTURE  EDIDC OPTIONAL
*"            TLINES STRUCTURE  TLINE OPTIONAL
*"----------------------------------------------------------------
* *** --- Reading the application Data --- ****
  CALL FUNCTION 'READ_TEXT'
      EXPORTING
          ID                    = T_HEAD-TDID
          LANGUAGE              = T_HEAD-TDSPRAS
          NAME                  = T_HEAD-TDNAME
          OBJECT                = T_HEAD-TDOBJECT
      IMPORTING
          HEADER                = E_THEAD
      TABLES
          LINES                 = TLINES.
* *** --- Packing the application data into IDoc
    MOVE E_THEAD TO IDOC_DATA-SDATA.
    MOVE 'YAXX_THEAD' TO IDOC_DATA-SEGNAM.
    APPEND IDOC_DATA.

    LOOP AT TLINES.
      MOVE E_THEAD TO IDOC_DATA-SDATA.
* *** -- we still need to fill more segment info
      MOVE 'YAXX_TLINE' TO IDOC_DATA-SEGNAM.
      APPEND IDOC_DATA.
    ENDLOOP.

* *** --- Packing the IDoc control record --- ****
  CLEAR IDOC_CONTRL.
  IDOC_CONTRL-IDOCTP = 'YAXX_TEXT'.
* *** -- we still should fill more control record info
  APPEND IDOC_CONTRL.

ENDFUNCTION.
```

Program 1:   Sample IDoc outbound function module

We will show a short example that packs SAP R/3 SAPscript standard text elements into IDocs and stores them back to texts in a second rotine. The text elements can be edited with SO10.

routine. The text elements can be edited with SO10.

Text from READ_TEXT
Each R/3 standard text elements has a header record which is stored in table STXH. The text lines itself are stored in a special cluster table. To retrieve the text from the cluster, you will use the standard function module function READ_TEXT.

Outbound processing
The program below will retrieve a text document from the text pool, convert the text lines into IDoc format and create the necessary control information.

Reading data
The first step is reading the data from the application database by calling the function module READ_TEXT.

```
* *** --- Reading the application Data --- ****
  CALL FUNCTION 'READ_TEXT'
      EXPORTING
            ID                      = T_HEAD-TDID
            LANGUAGE                = T_HEAD-TDSPRAS
            NAME                    = T_HEAD-TDNAME
            OBJECT                  = T_HEAD-TDOBJECT
      IMPORTING
            HEADER                  = E_THEAD
      TABLES
            LINES                   = TLINES.
```

Converting application data into IDoc format
Our next duty is to pack the data into the IDoc record. This means moving the application data to the data part of the IDoc record structure EDIDD and fill the corresponding segment information.

```
* *** --- Packing the application data into IDoc
     MOVE E_THEAD TO IDOC_DATA-SDATA.
*     the receiver needs the segment name
     in order to interpret the segment
     MOVE 'YAXX_THEAD' TO IDOC_DATA-SEGNAM.
     APPEND IDOC_DATA.

     LOOP AT TLINES.
       MOVE E_THEAD TO IDOC_DATA-SDATA.
* *** -- we still need to fill more segment info
       MOVE 'YAXX_TLINE' TO IDOC_DATA-SEGNAM.
         APPEND IDOC_DATA.
     ENDLOOP.
```

Filling control record information
Finally we have to provide a correctly filled control record for this IDoc. If the IDoc routine is used in a standard automated environment, it is usually sufficient to fill the field EDIDC-IDOCTP with the IDoc type, EDIDC-MESTYP with the context message type and the receiver name. The remaining fields are automatically filled by the standard processing routines if applicable.

```
* *** --- Packing the IDoc control record --- ****
  CLEAR IDOC_CONTRL.
  IDOC_CONTRL-IDOCTP = 'YAXX_TEXT'.
* *** -- we still need to fill more control rec info
  APPEND IDOC_CONTRL.
```

5.3   Sample Inbound Routines

Inbound processing is widely the reverse process of an outbound.. The received IDoc has to be unpacked, interpreted and transferred to an application for further processing.

```
FUNCTION
*"----------------------------------------------------------------------
*"*"Lokale Schnittstelle:
*"      IMPORTING
*"            VALUE(INPUT_METHOD) LIKE  BDWFAP_PAR-INPUTMETHD
*"            VALUE(MASS_PROCESSING) LIKE  BDWFAP_PAR-MASS_PROC
*"      EXPORTING
*"            VALUE(WORKFLOW_RESULT) LIKE  BDWFAP_PAR-RESULT
*"            VALUE(APPLICATION_VARIABLE) LIKE  BDWFAP_PAR-APPL_VAR
*"            VALUE(IN_UPDATE_TASK) LIKE  BDWFAP_PAR-UPDATETASK
*"            VALUE(CALL_TRANSACTION_DONE) LIKE  BDWFAP_PAR-CALLTRANS
*"      TABLES
*"            IDOC_CONTRL STRUCTURE  EDIDC
*"            IDOC_DATA STRUCTURE  EDIDD
*"            IDOC_STATUS STRUCTURE  BDIDOCSTAT
*"            RETURN_VARIABLES STRUCTURE  BDWFRETVAR
*"            SERIALIZATION_INFO STRUCTURE  BDI_SER
*"----------------------------------------------------------------------
  DATA: XTHEAD    LIKE THEAD   .
  DATA: TLINES LIKE TLINE    OCCURS 0 WITH HEADER LINE.

  CLEAR XTHEAD.
  REFRESH TLINES.

* *** --- Unpacking the IDoc --- ***
  LOOP AT IDOC_DATA.
    CASE IDOC_DATA-SEGNAM.
      WHEN 'YAXX_THEAD'.
           MOVE IDOC_DATA-SDATA TO XTHEAD.
      WHEN 'YAXX_TLINE'.
           MOVE IDOC_DATA-SDATA TO TLINES.
    ENDCASE.
  ENDLOOP.

* *** --- Calling the application to process the received data --- ***
  CALL FUNCTION 'SAVE_TEXT'
      EXPORTING
          HEADER         = XTHEAD
          SAVEMODE_DIRECT = 'X'
      TABLES
          LINES          = TLINES.

  ADD SY-SUBRC TO OK.
* füllen IDOC_Status
* fill IDOC_Status
    IDOC_STATUS-DOCNUM = IDOC_CONTRL-DOCNUM.
    IDOC_STATUS-MSGV1  = IDOC_CONTRL-IDOCTP.
    IDOC_STATUS-MSGV2  = XTHEAD.
    IDOC_STATUS-MSGID  = '38'.
    IDOC_STATUS-MSGNO  = '000'.
    IF OK NE 0.
      IDOC_STATUS-STATUS = '51'.
      IDOC_STATUS-MSGTY  = 'E'.
```

```
    ELSE.
       IDOC_STATUS-STATUS = '53'.
       IDOC_STATUS-MSGTY  = 'S'.
       CALL_TRANSACTION_DONE = 'X'.
    ENDIF.
    APPEND IDOC_STATUS.
ENDFUNCTION.
```

Program 2:   Sample IDoc outbound function module

Inbound processing function module
This example of a simple inbound function module expects an IDoc with rows of plain text as created in the outbound example above. The procedure will extract the text name and the text line from the IDoc and hand over the text data to the function module SAVE_TEXT which will store the text in the text pool.

Unpacking the IDoc data The received IDoc data is processed record by record and data is sorted out according the segment type.

```
* *** --- Unpacking the IDoc --- ***
  LOOP AT IDOC_DATA.bb
    CASE IDOC_DATA-SEGNAM.
      WHEN 'YAXX_THEAD'.
          PERFORM UNPACK_IDOC TABLES IDOC_DATA USING XTHEAD.
      WHEN 'YAXX_TLINE'.
          PERFORM UNPACK_TAB  TABLES IDOC_DATA TLINES.
    ENDCASE.
  ENDLOOP.
```

Storing data
When the IDoc is unpacked data is passed to the application.

```
* *** --- Calling the application to process the received data --- ***
  CALL FUNCTION 'SAVE_TEXT'
      EXPORTING
           HEADER        = XTHEAD
      TABLES
           LINES         = TLINES.
```

Writing a status log
Finally the processing routine needs to pass a status record to the IDoc processor. This status indicates successful or unsuccessful processing and will be added as a log entry to the table EDIDS.

```
* fill IDOC_Status
    IF OK NE 0.
       IDOC_STATUS-STATUS = '51'.
*      IDOC_STATUS-.. = . fill the other fields to log information
    ELSE.
       IDOC_STATUS-STATUS = '53'.
    ENDIF.
    APPEND IDOC_STATUS.
```

The status value '51' indicates a general error during application processing and the status '53' indicates everything is OK.

6

IDocs Terminology And Basic Tools

This is a collection of expressions used in context with IDocs. You should be familiar with them. Some are also used in non-IDoc context with a completely different meaning, e.g. the term *message*, so avoid misunderstandings. Many fights in project teams arise from different interpretations of the same expression.

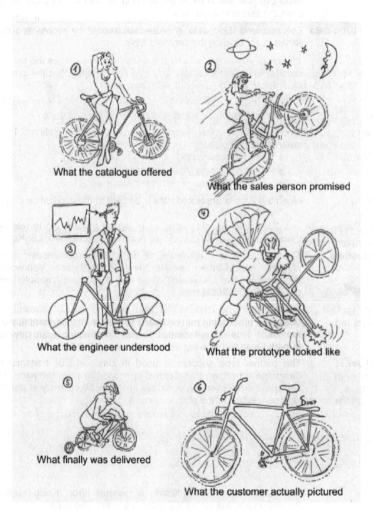

What the catalogue offered

What the sales person promised

What the engineer understood

What the prototype looked like

What finally was delivered

What the customer actually pictured

6.1 Basic Terms

There are a couple of expressions and methods that you need to know, when dealing with IDoc.

Message Type

The message type defines the semantic context of an IDoc. The message type tells the processing routines, how the message has to be interpreted.

The same IDoc data can be sent with different message types. E.g. the same IDoc structure which is used for a purchase order can also be used for transmitting a sales order. Imagine the situation that you receive a sales order from your clients and in addition you receive copies of sales orders sent by an subsidiary of your company.

IDoc Type

An IDoc type defines the syntax of the IDoc data. It tells which segments are found in an IDoc and what fields the segments are made of.

Processing Code

The processing code is a logical name that determines the processing routine. This points usually to a function module, but the processing routine can also be a workflow or an event.

The use of a logical processing code makes it easy to modify the processing routine for a series of partner profiles at once.

Partner profile

Every sender-receiver relationship needs a profile defined. This one determines
- the processing code
- the processing times and conditions
- and in the case of outbound IDocs also
- the media port used to send the IDoc and
- the triggers used to send the IDoc

Partner Type

The IDoc partners are classified in logical groups. Up to release 4.5 there were the following standard partner types defined: LS, KU, LI.

LS - Logical Systems

The logical system is meant to be a different computer and was primarily introduced for use with the ALE functionality. You would use a partner type of LS, when linking with a different computer system, e.g. a legacy or subsystem.

KU - Customer [ger. Kunde]

The partner type customer is used in classical EDI transmission to designate a partner, that requires a service from your company or is in the role of a debtor with respect to your company, e.g. the payer, sold-to-party, ship-to-party.

LI - Supplier [Ger.: Lieferant]

The partner type supplier is used in classical EDI transmission to designate a partner, that delivers a service to your company. This is typically the supplier in a purchase order. In SD orders you also find LI type partners, e.g. the shipping agent.

6.2   Terminology

6.2.1   Message Type – How to Know What the Data Means

Data exchanged by an IDoc via EDI is known as message. Messages of the same kind belong to the same message type.

Define the semantic context	The message type defines the semantic context of an IDoc. The message type tells the receiver, how the message has to be interpreted.
Messages are information for a foreign partner	The term message is commonly used in communication, be it EDI or telecommunication. Any stream of data sent to a receiver with a well-defined information in it, is known as a message. EDIFACT, ANSI/X.12, XML and others use message the same way.
The term message is also used for R/3's internal communication between applications	Unfortunately, the term message is used in many contexts other than EDI as well. Even R/3 uses the word message for the internal communication between applications. While this is totally OK from the abstract point of view of data modelling, it may sometimes cause confusion, if it is unclear whether we talk about IDoc messages or internal messages.
	The specification of the message type along with the sent IDoc package is especially important, when the physical IDoc type (the data structure of the IDoc file) is used for different purposes.
	A classical ambiguity arises in communication with customs via EDI. The usually set up a universal file format for an arbitrary kind of declarations, e.g. Intrastat, Extrastat, Export declarations, monthly reports etc. Depending on the message type, only applicable fields are filled with valid data. The message type tells the receiver, which fields are of interest at all.

6.2.2   Partner Profiles – How to Know the Format of the Partner

Different partners may speak different languages. While the information remains the same, different receivers may require completely different file formats and communication protocols. This information is stored in a partner profile.

Partner Profiles are the catalogue of active EDI connection from and to R/3	In a partner profile you will specify the names of the partners which are allowed to exchange IDocs to your system. For each partner you have to list the message types which the partner may send.
Partner profiles stores the IDoc type to use	For any such message type, the profile tells the IDoc type, which the partner expects for that kind of message.

Outbound customizing agrees how data is electronically exchanged	For outbound processing, the partner profile also sets the media to transport the data to its receiver, e.g. • an operating system file • automated FTP • XML or EDIFACT transmission via a broker/converter • internet • direct remote function call The mean of transport depends on the receiving partner, the IDoc type and message type (context).
Different partners, different profiles	So you may determine to send the same data as a file to your vendor and via FTP to your remote plant. Also you may decide to exchange purchase data with a vendor via FTP but send payment notes to the same vendor in a file.
Inbound customizing determines the processing routine	For inbound processing, the partner profile customizing will also determine a processing code, which can handle the received data. The partner profile may tell you the following:

- Supplier .... MAK_CO
  sends the message .... SHIPPING_ADVISE
  via the port named.......................... INTERNET
  using IDoc type................................ SHPADV01
  processed with code......................... SHIPMENTLEFT

- Sales agent .... LOWSELL
  sends the message .... SALESORDERS
  via the port named.......................... RFCLINK
  using IDoc type................................ ORDERS01
  processed with code......................... CUSTOMERORDER

- Sales agent .... SUPERSELL
  sends the message .... SALESORDERS
  via the port named.......................... RFCLINK
  using IDoc type................................ ORDERS01
  processed with code......................... AGENTORDER

6.2.3   IDoc Type – The Structure of The IDoc File

The IDoc type is the name of the data structure used to describe the file format of a specific IDoc.

IDoc type defines the structure of the segments	An IDoc is a segmented data file. It has typically several segments. The segments are usually structured into fields, however different segments use different fields. The IDoc type is defined with transaction *WE30*, the respective segments are defined with transaction *WE31*.

6.2.4   Processing Codes

The processing code is a pointer to an algorithm to process an IDoc. It is used to allow more flexibility in assigning the processing function to an IDoc message.

The logical processing code determines the algorithm in R/3 used to process the IDoc

The processing code is a logical name for the algorithm used to process the IDoc. The processing code points itself to a method or function, which is capable of processing the IDoc data.

A processing code can point to an SAP predefined or a self-written business object or function module as long as they comply with certain interface standards.

Allows to change the algorithm easily

The processing codes allow to easily change the processing algorithm. Because the process code can be used for more than one partner profile, the algorithm can be easily changed for every concerned IDoc.

The processing code defines a method or function to process an IDoc

The IDoc engine will call a function module or a business object which is expected to perform the application processing for the received IDoc data. The function module must provide exactly the interface parameters which are needed to call it from the IDoc engine.

7

IDocs Customizing

Let aside the writing of the processing function modules, IDoc development requires the definition of the segment structures and a series customizing settings to control the flow of the IDoc engine.

Summary

- Customize basic installation parameters
- Define segment structures
- Define message types, processing codes

7.1 Basic Customizing Settings

Segments define the structure of the records in an IDoc. They are defined with transaction WE31.

Check first, whether the client you are working in, has already a logical system name assigned.

T000 – name of own logical system

The logical system name is stored in table `T000` as `T000-LOGSYS`. This is the table of installed clients.

TBDLS – list of known logical destinations

If there is no name defined, yet, you need to create a logical system name before. This means simply adding a line to table `TBDLS`. You can edit the table directly or access the table from transaction `SALE`.

Naming conventions:
DEVCLNT100
PROCLNT123
TSTCLNT999

The recommended naming convention is

`sysid + "CLNT" + client`

If your system is `DEV` and client `100`, then the logical system name should be: `DEVCLNT100`.

System `PRO` with client `123` would be `PROCLNT123` etc.

SM59 – define physical destination and characteristics of a logical system

The logical system needs also be defined as a target within the R/3 network. Those definitions are done with transaction SM59 and are usually part of the work of the R/3 basis team.

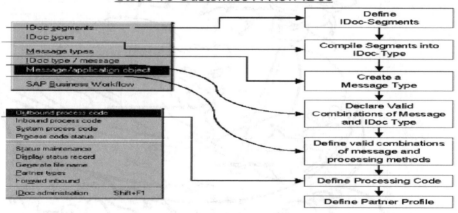

Steps To Customise A New IDoc

Illustration 8:   **Step to customize outbound IDoc processing**

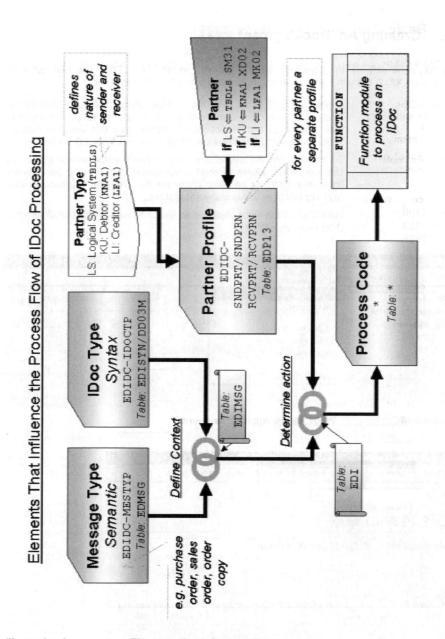

Illustration 9: **Elements that influence IDoc processing**

7.2 Creating An IDoc Segment `WE31`

The segment defines the structure of the records in an IDoc. They are defined with transaction `WE31`. We will define a structure to send a text from the text database.

Define a DDic structure with `WE31`

Transaction `WE31` calls the IDoc segment editor. The editor defines the fields of a single segment structure. The thus defined IDoc segment is then created as a data dictionary structure. You can view the created structure with `SE11` and use it in an ABAP as any TABLES declaration.

Example:

To demonstrate the use of the IDoc segment editor we will set up an example, which allows to send a single text from the text pool (tables `STXH` and `STXL`) as an IDoc. These are the texts that you can see with SO10 or edit from within many applications.

We will show the steps to define an IDoc segment `YAXX_THEAD` with the DDic structure of `THEAD`.

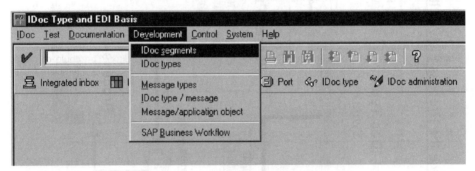

Illustration 1:   **WE31, define the IDoc segment**

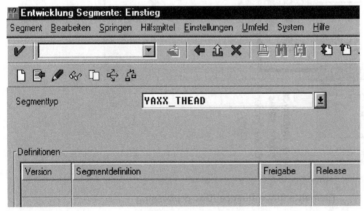

Illustration 2:   **Naming the segment**

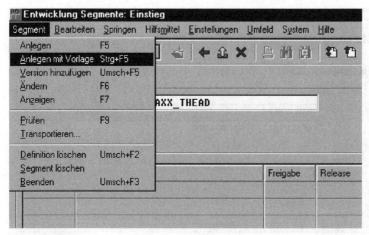

Illustration 3:   Selecting a template

Copy the segment structure from a DDic object

To facilitate our work, we will use the "copy-from-template-tool", which reads the definition of a DDIC structure and inserts the field and the matching definitions as rows in the IDoc editor. You could of course define the structure completely manually, but using the template makes it easier.

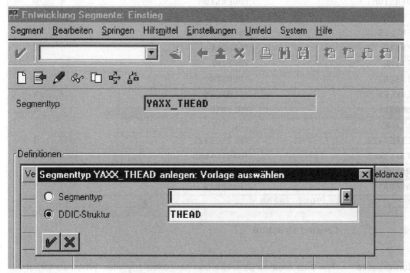

Illustration 4:   Now select it really

The tool in release 4.0b lets you to use both DDIC structures or another IDoc segment definition as a template.

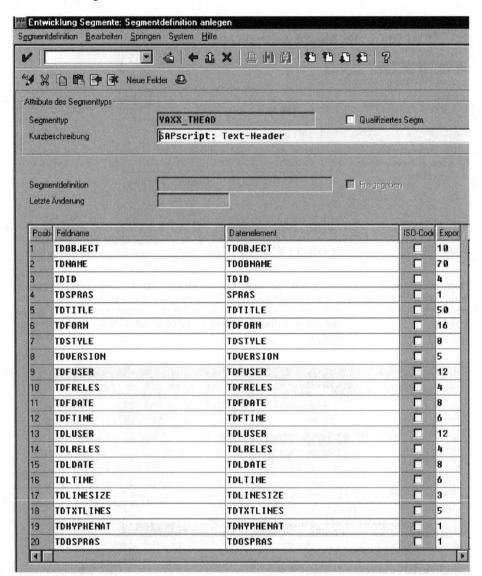

IDocs Customizing

Illustration 5: **Created structure**

The definition creates automatically a corresponding DDic structure

The thus created structure can be edited any time. When saving, it will create a data dictionary structure based on the definition in WE31. The DDIC structure will retain the same name. You can view the structure as a table definition with `SE11` and use it in an ABAP the same way.

7.3 Defining The Message Type (EDMSG)

The message type defines the context under which an IDoc is transferred to its destination. It allows to use the same IDoc file format to use for several different applications.

Sales order becomes purchase order for receiver	Imagine the situation of sending a purchase order to a supplier. When the IDoc with the purchase order reaches the supplier, it will be interpreted as a sales order received from a customer, namely you.
Sales order can be forwarded and remains a sales order	Simultaneously you want to send the IDoc data to the suppliers warehouse to inform it, that a purchase order has been issued and is on the way.
	Both IDoc receivers will receive the same IDoc format, however the IDoc will be tagged with a different message type. While the IDoc to the supplier will be flagged as a purchase order (in SAP R/3 standard: message type = ORDERS), the same IDoc sent to the warehouse should be flagged differently, so that the warehouse can recognize the order as a mere informational copy and process them differently than a true purchase order.
Message type plus IDoc type determine processing algorithm	The message type together with the IDoc type determine the processing function.
EDMSG	The message types are stored in table EDMSG.
WEDI	Defining the message type can be done from the transaction WEDI

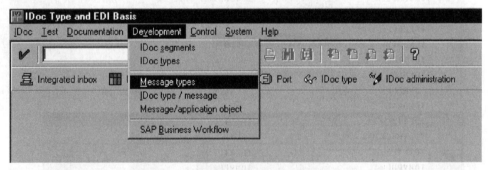

Illustration 6:	**EDMSG: Defining The Message Type (1)**

EDMSG used as check table	The entry is only a base entry which tells the system, that the message type is allowed. Other transactions will use that table as a check table to validate the entry.

http://idocs.de http://logosworld.de

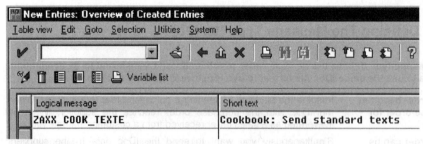

Illustration 7:   EDMSG: Defining The Message Type (2)

7.4 Define Valid Combination Of Message and IDoc Types

The valid combinations of message type and IDoc type are stored in table EDIMSG.

Used for validation   The declaration of valid combinations is done to allow validation, if the system can handle a certain combination.

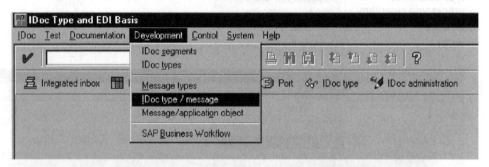

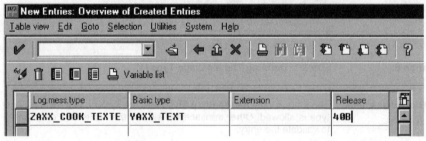

Illustration 8:   **EDIMSG: Define Valid Combination Of Message and IDoc Types**

7.5 Assigning a processing function (Table `EDIFCT`)

The combination of message type and IDoc type determine the processing algorithm. This is usually a function module with a well defined interface or a SAP business object and is set up in table EDIFCT.

The entry made here points to a function module, which will be called when the IDoc is processed.

The entries for message code and message function are usually left blank. They can be used to derive sub types of messages together with the partner profile used.

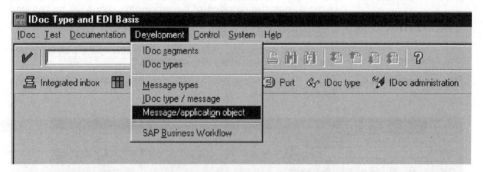

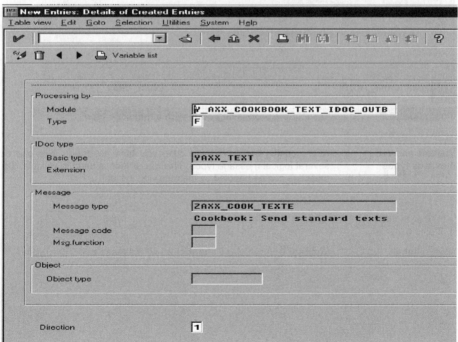

Illustration 9:   **Assign a handler function to a message/message type**

7.6   Processing Codes

R/3 uses the method of logical process codes to detach the IDoc processing and the processing function module. They assign a logical name to function instead of specifying the physical function name.

Logical pointer to a processing method
The IDoc functions are often used for a serious of message type/IDoc type combination. It happens that you need to replace the processing function by a different one. E.g. when you make a copy of a standard function to avoid modifying the standard.

Easy replacing of the processing method
The combination message type/IDoc will determine the logical processing code, which itself points to a function. If the function changes, only the definition of the processing codes will be changed and the new function will be immediately effective for all IDocs associated with the process code.

For inbound processing codes you have to specify the method to use for the determination of the inbound function.

Processing with ALE
This is the option you would usually choose. It allows processing via the ALE scenarios.

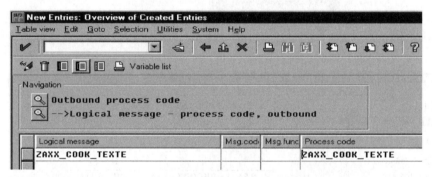

Illustration 10:   **Associate a processing code with a message type**

Validate allowed message types
After defining the processing code you have to assign it to one or several logical message types. This declaration is used to validate, if a message can be handled by the receiving system.

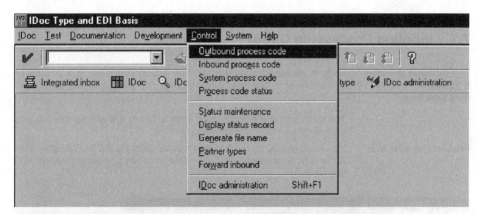

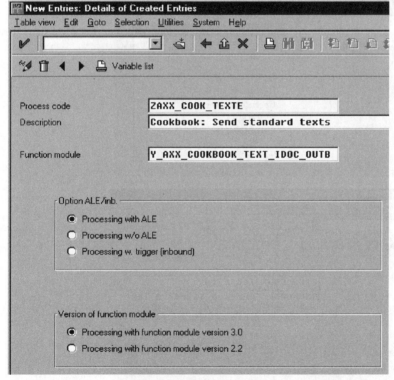

Illustration 11:   **Assign an outbound processing code (Step 1)**

7.7   Inbound Processing Code

The inbound processing code is assigned analogously. The processing code is a pointer to a function module which can handle the inbound request for the specified IDoc and message type.

The definition of the processing code is telling the handler routine and assigning a serious of processing options.

Processing with ALE   You need to tick, if your function can be used via the ALE engine. This is the option you would usually choose. It allows processing via the ALE scenarios.

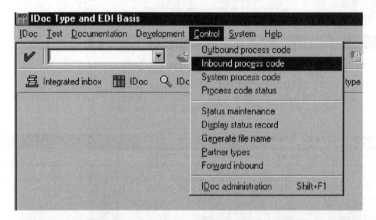

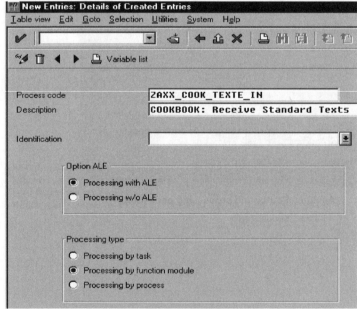

Associate a function module with a process code

Table TBD51 to define if visible BTCI is allowed

For inbound processing you need to tell. whether the function will be capable of dialog processing. This is meant for those functions, which process the inbound data via call transaction. Those functions can be replayed in visible batch input mode to check why the processing might have failed.

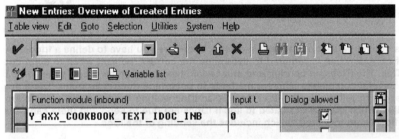

Illustration 12: **Define if the processing can be done in dialog via call transaction**

Validate allowed message types

After defining the processing code you have to assign it to one or several logical message types. This declaration is used to validate, if a message can be handled by the receiving system.

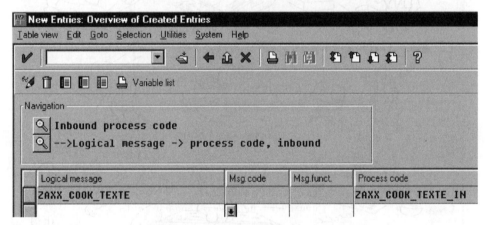

Illustration 13: **Associate a processing code with a message type**

The examples above showed only the association with a function module. You can also define business objects with transaction *SWO1* and define them as a handler. For those familiar with the object model of R/3 it may be a design decision. In this book, we will deal with the function modules only.

http://idocs.de http://logosworld.de

8

IDoc Outbound Triggers

IDocs should be sent out at certain events. Therefore you have to define a trigger. A lot of consideration is required to determine the correct moment when to send out the IDoc. The IDoc can be triggered at a certain time or when an event is raised. R/3 uses several completely different methods to determine the trigger point. There are messages to tell the system that there is an IDoc waiting for dispatching, there are log files which may be evaluated to see if IDocs are due to send and there can be a workflow chain triggered, which includes the sending of the IDoc.

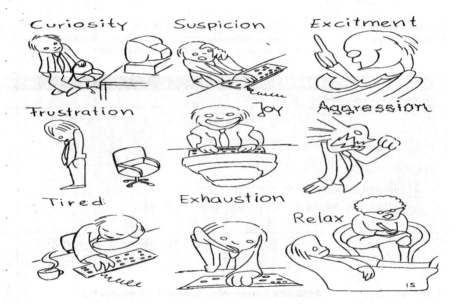

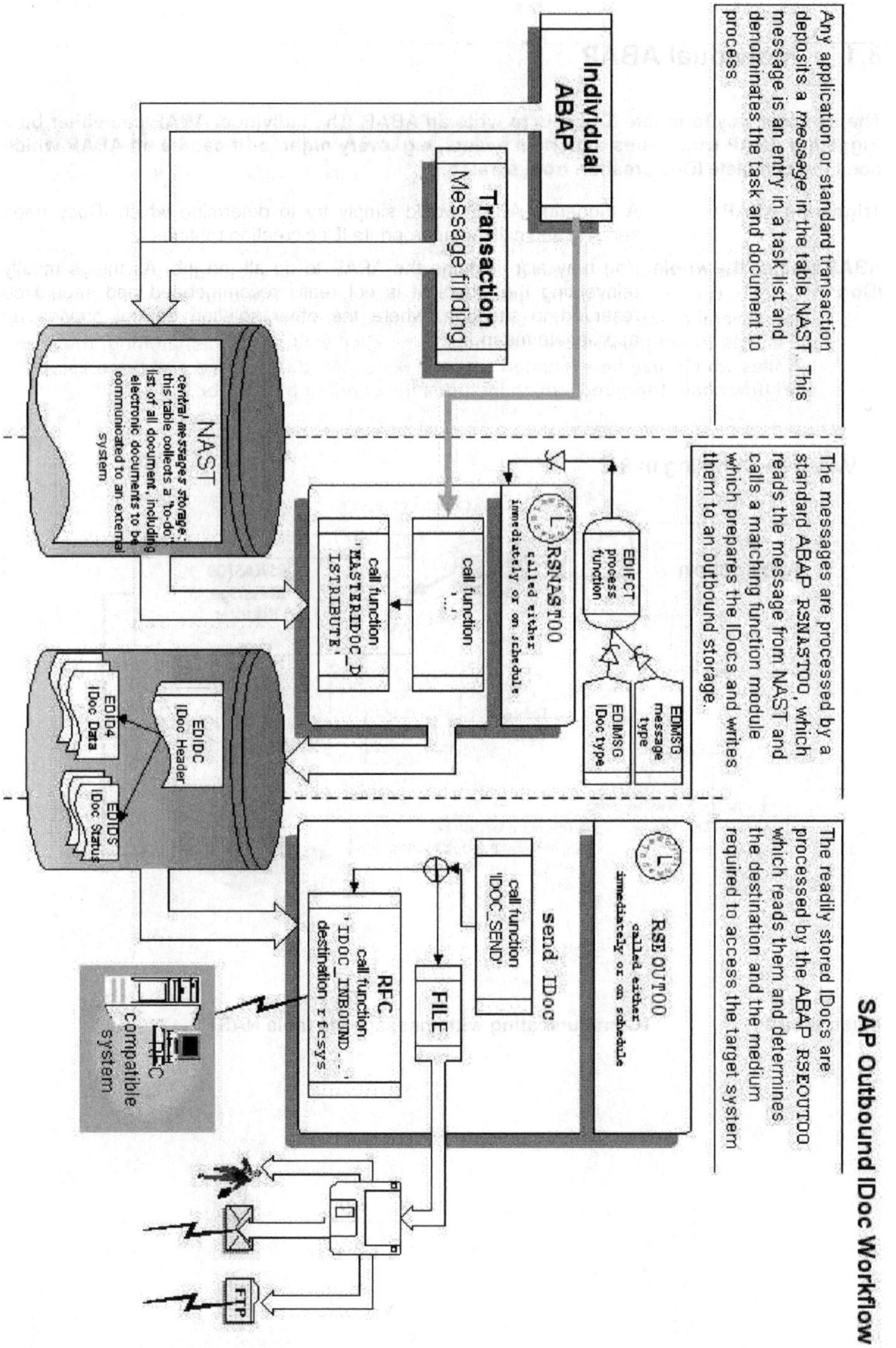

IDoc Outbound Triggers

Any application or standard transaction deposits a *message* in the table NAST. This message is an entry in a task list and denominates the task and document to process.

Individual ABAP

Transaction
Message finding

NAST
central messages storage; this table collects a 'to-do' list of all document, including electronic document, to be communicated to an external system

SAP Outbound IDoc Workflow

The messages are processed by a standard ABAP RSNAST00, which reads the message from NAST and calls a matching function module which prepares the message the IDocs and writes them to an outbound storage.

L RSNAST00
called either immediately or on schedule

process function

EDIFCT

call function
'MASTERIDOC_D ISTRIBUTE'

call function ...

EDIMSG
message type

EDIMSG
IDoc type

EDIDC
IDoc Header

EDID4
IDoc Data

EDIDS
IDoc Status

The readily stored IDocs are processed by the ABAP RSEOUT00, which reads them and determines the destination and the medium required to access the target system

L RSEOUT00
called either immediately or on schedule

send IDoc

call function
'IDOC_SEND'

FILE

RFC
call function
'IDOC_INBOUND...'
destination r:cssys

RFC compatible system

FTP

Illustration 10: **General Process logic of IDoc outbound**

8.1 Individual ABAP

The simplest way to create IDocs, is to write an ABAP. The individual ABAP can either be a triggering ABAP which runs at certain events, e.g. every night, or it can be an ABAP which does the complete IDoc creation from scratch.

Triggering ABAP   A triggering ABAP would simply try to determine which IDocs need sending and call the appropriate IDoc creation routines.

ABAP creates the whole IDoc   You may also imagine the ABAP to do all the job. As this is mostly reinventing the wheel, it is not really recommended and should be reserved to situation, where the other solution do not provide an appropriate mean.

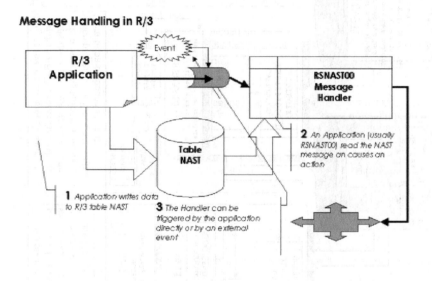

Illustration 11:   Communicating with message via table NAST

8.2   NAST Messages Based Outbound IDocs

You can use the R/3 message concept to trigger IDocs the same way as you trigger SAPscript printing.

One of the key tables in R/3 is the table NAST. This table records reminders written by applications. Those reminders are called messages.

Applications write messages to NAST, which will be processed by a message handler

Every time when an applications sees the necessity to pass information to a third party. a message is written to NAST. A message handler will eventually check the entries in the table and cause an appropriate action.

EDI uses the same mechanism as printing

The concept of NAST messages has originally been designed for triggering SAPscript printing. The very same mechanism is used for IDocs, where the IDoc processor replaces the print task, as an IDoc is only the paperless form of a printed document.

Condition technique can mostly be used

The messages are usually be created using the condition technique, a mechanism available to all major R/3 applications.

Printing, EDI and ALE use the same trigger

The conditions are set up the same way for any output media. So you may define a condition for printing a document and then just change the output media from printer to IDoc/EDI or ALE.

NAST messages are created by application by calling function module MESSAGING

Creating NAST messages is a standard functionality in most of the SAP core applications. Those applications - e.g. VA01, ME21 - perform calls to the central function module MESSAGING of group V61B. The function module uses customizing entries, mainly those of the tables T681* to T685*.

NAST contains object key, sender and receiver

A NAST output message is stored as a single record in the table NAST. The record stores all information that is necessary to create an IDoc. This includes mainly an object key to identify the processed object and application to the message handler and the sender and receiver information.

Programs RSNAST00 and RSNASTED provide versatile subroutines for NAST processing

The messages are typically processed by
FORM ENTRY in PROGRAM RSNAST00.
If we are dealing with printing or faxing and
FORM EDI_PROCESSING in PROGRAM RSNASTED.
If we are dealing with IDocs
FORM ALE_PROCESSING in PROGRAM RSNASTED.
If we are dealing with ALE.

The following piece of code does principally the same thing as RSNAST00 does and makes full use of all customizing settings for message handling.

FORM einzelnachricht IN PROGRAM RSNAST00

TABLES: NAST.
SELECT * FROM NAST ...
PERFORM einzelnachricht IN PROGRAM RSNAST00

Programs are customized in table TNAPR

The processing routine for the respective media and message is customized in the table TNAPR. This table records the name of a FORM routine, which processes the message for the chosen media and the name of an ABAP where this FORM is found.

8.3   The RSNAST00 ABAP

The ABAP RSNAST00 is the standard ABAP, which is used to collect unprocessed NAST message and to execute the assigned action.

RSNAST00 is the standard batch collector for messages

RSNAST00 can be executed as a collector batch run, that eventually looks for unprocessed IDocs. The usual way of doing that is to define a batch-run job with transaction *SM37*. This job has to be set for periodic processing and start a program that triggers the IDoc re-sending.

RSNAST00 processes only messages of a certain status

Cave! RSNAST00 will only look for IDocs which are set to NAST-VSZTP = '1' or '2' (Time of processing). VSZPT = '3' or '4' is ignored by RSNAST00.

For batch execution a selection variant is required

Start RSNAST00 in the foreground first and find the parameters that match your required selection criteria. Save them as a VARIANT and then define the periodic batch job using the variant.

If RSNAST00 does not meet 100% your needs you can create an own program similar to RSNAST00. The only requirement for this program are two steps:

```
* Read the NAST entry to process into structure NAST
tables nast.
data: subrc like sy-subrc.....
select from NAST where .......
* then call FORM einzelnachricht(rsnast00) to
process the record
PERFORM einzelnachricht(rsnast00) USING subrc.
```

8.4   Sending IDocs Via RSNASTED

Standard R/3 provides you with powerful routines, to trigger, prepare and send out IDocs in a controlled way. There are only a few rare cases, where you do not want to send IDocs the standard way.

The ABAP `RSNAST00` is the standard routine to send IDocs from entries in the message control. This program can be called directly, from a batch routine with variant or you can call the `FORM einzelnachricht_screen(RSNAST00)` from any other program, while having the structure NAST correctly filled with all necessary information.

RSNAST00 determines if it is IDoc or SAPscript etc.

If there is an entry in table `NAST`, `RSNAST00` looks up the associated processing routine in table `TNAPR`. If it is to send an IDoc with standard means, this will usually be the routine `RSNASTED(EDI_PROCESSING)` or `RSNASTED(ALE_PROCESSING)` in the case of ALE distribution.

RSNASTED processes IDocs

RSNASTED itself determines the associated IDoc outbound function module, executes it to fill the EDIDx tables and passes the prepared IDoc to the port.

You can call the standard processing routines from any ABAP, by executing the following call to the routine. You only have to make sure that the structure NAST is declared with the tables statement in the calling routine and that you fill at least the key part and the routine (TNAPR) information before.

```
TABLES NAST.
NAST-MANDT = SY-MANDT.
NAST-KSCHL = 'ZEDIK'.
NAST-KAPPL = 'V1'.
NAST-OBJKY = '0012345678'.
NAST-PARNR = 'D012345678'.
PERFORM einzelnachricht_screen(RSNAST00).
```

Calling `einzelnachricht_screen` determines how the message is processed. If you want to force the IDoc-processing you can call it directly:

```
TNAPR-PROGN = ''.
TNAPR-ROUTN = 'ENTRY'.
PERFORM edi_processing(RSNASTED).
```

8.5   Sending IDocs Via RSNAST00

Here is the principle flow how RSNAST00 processes messages for IDocs.

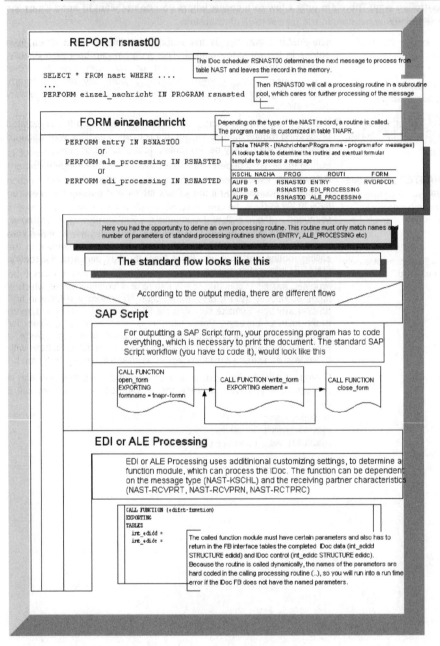

Illustration 12:   **Process logic of RSNAST00 ABAP**

8.6   Workflow Based Outbound IDocs

Unfortunately, there are application that do not create messages. This is especially true for master data applications. However, most applications fire a workflow event during update, which can easily be used to trigger the IDoc distribution.

`SWE_EVENT_CREATE`	Many SAP R/3 applications issue a call to the function `SWE_EVENT_CREATE` during update. This function module ignites a simple workflow event.
Workflow is a call to a function module	Technically a workflow event is a timed call to a function module, which takes the issuing event as the key to process a subsequent action.
Applications with change documents always trigger workflow events	If an application writes regular change documents (*ger.: Änderungsbelege*) to the database, it will issue automatically a workflow event. This event is triggered from within the function `CHANGEDOCUMENT_CLOSE`. The change document workflow event is always triggered, independent of the case whether a change document is actually written.
Workflow coupling can be done by utility functions	In order to make use of the workflow for IDoc processing, you do not have to go through the cumbersome workflow design procedure as it is described in the workflow documentation. For the mentioned purpose, you can register the workflow handler from the menu, which says Event Coupling from the *BALD* transaction.
Workflow cannot easily be restarted	Triggering the IDoc from a workflow event has a disadvantage: if the IDoc has to be repeated for some reason, the event cannot be repeated easily. This is due to the nature of a workflow event, which is triggered usually from a precedent action. Therefore you have to find an own way how to make sure that the IDoc is actually generated, even in the case of an error. Practically this is not a very big problem for IDocs. In most cases the creation of the IDoc will always take place. If there is a problem, then the IDoc would be stored in the IDoc base with a respective status. It will shown in transaction *WE05* and can be resend from there.

8.7 Workflow Event From Change Document

Instead of waiting for a polling job to create IDocs, they can also be created immediately after a transaction finishes. This can be done by assigning an action to an workflow event.

Workflow events are usually fired from an update routine	Most application fire a workflow event from the update routine by calling the function

```
FUNCTION swe_event_create
```

SWLD lets install and log workflows	You can check if an application fires events by activating the event log from transaction *SWLD*. Calling and saving a transaction will write the event's name and circumstances into the log file.

If an application does not fire workflow events directly, there is still another chance that a workflow may be used without touching the R/3 original programs.

Workflow Events are also fired from change document	Every application that writes change documents triggers a workflow event from within the function module CHANGEDOCUMENT_CLOSE, which is called form the update processing upon writing the change document. This will call the workflow processor

```
FUNCTION swe_event_create_changedocument
```

Both workflow types are not compatible with each other with respect to the function modules used to handle the event.

The workflow types are incompatible but work according the same principal	Both will call a function module whose name they find in the workflow linkage tables. swe_event_create will look in table SWETYPECOU while swe_event_create_changedocument would look in SWECDOBJ for the name of the function module.
The workflow handler will be called dynamically	If a name is found, the function module will then be called dynamically. This is all to say about the linkage of the workflow.

The dynamic call looks like the following.

```
CALL FUNCTION swecdobj-objtypefb
EXPORTING
changedocument_header = changedocument_header
objecttype = swecdobj-objtype
IMPORTING
objecttype = swecdobj-objtype
TABLES
changedocument_position = changedocument_position.
```

8.8   ALE Change Pointers

Applications which write change documents will also try to write change pointers for ALE operations. These are log entries to remember all modified data records relevant for ALE.

Most applications write change documents. These are primarily log entries in the tables CDHDR and CDPOS.

Change docs remember changes in transaction

Change documents remember the modified fields made to the database by an application. They also remember the user name and the time when the modification took place.

Data elements are marked to be relevant for change documents

The decision whether a field modification is relevant for a change document is triggered by a flag of the modified field's data element. You can set the flag with *SE11* by modifying the data element.

ALE may need other triggers

For the purpose of distributing data via ALE to other systems, you may want to choose other fields, which shall be regarded relevant for triggering a distribution.

Therefore R/3 introduced the concept of change pointers, which are nothing else than a second log file specially designed for writing the change pointers which are meant to trigger IDoc distribution via ALE.

Change pointers remember key of the document

So the change pointers will remember the key of the document every time when a relevant field has changed.

An ABAP creates the IDocs

Change pointers are then evaluated by an ABAP which calls the IDoc creation, for every modified document found in the change pointers.

Change pointers are when change documents have been written

The Change pointers are written from the routine CHANGEDOCUMENT_CLOSE when saving the generated change document. So change pointers are automatically written when a relevant document changes.

The following function is called from within CHANGEDOCUMENT_CLOSE in order to write the change pointers.

```
CALL FUNCTION 'CHANGE_POINTERS_CREATE'
EXPORTING
change_document_header = cdhdr
TABLES
change_document_position = ins_cdpos.
```

8.9   Activation of change pointer update

Change pointers are log entries to table BDCP which are written every time a transaction modifies certain fields. The change pointers are designed for ALE distribution and written by the function CHANGE_DOCUMENT_CLOSE.

Change pointers are written for use with ALE. There are ABAPs like RBDMIDOC which can read the change pointers and trigger an IDoc for ALE distribution.

The change pointers are mainly the same as change documents. They however can be set up differently, so fields which trigger change documents are not necessarily the same that cause change pointers to be written.

In order to work with change pointers there are two steps to be performed

1. Turn on change pointer update generally
2. Decide which message types shall be included for change pointer update

Activate Change Pointer Generally

R3 allows to activate or deactivate the change pointer update. For this purpose it maintains a table TBDA1. The decision whether the change pointer update is active is done with a

Function Ale_Component_Check

Currently (release 40B) this check does nothing else than to check, if this table has an entry or not. If there is an entry in TBDA1, the ALE change pointers are generally active. If this table is empty, change pointers are turned off for everybody and everything, regardless of the other settings.

The two points read like you had the choice between turning it on generally or selectively. This is not the case: you always turn them on selectively. The switch to turn on generally is meant to activate or deactivate the whole mechanism.

reading the change pointers which are not yet processed

The change pointers which have not been processed yet, can be read with a function module.

Call Function 'CHANGE_POINTERS_READ'

RBDMIDOC

The ABAP RBDMIDOC will process all open change pointers and distribute the matching IDocs.

Use Change Documents Instead Of Change Pointers

When you want to send out an IDoc unconditionally every time a transaction updates, you better use the workflow from the change documents.

8.10 Dispatching ALE IDocs for Change Pointers

Change pointers must be processed by an ABAP, e.g. RBDMIDOC.

RBDMIDOC processes change pointers and sends the IDocs

The actual distribution of documents from change pointers must be done by an ABAP, which reads the change pointers and processes them. The standard ABAP for that is `RBDMIDOC`. For recurring execution it can be submitted in a scheduled job using *SM35* .

Function module defined in table TBDME

It then calls dynamically a function module whose name is stored in table `TBDME` for each message type.

```
Call Function Tbdme-IDocfbname
    Exporting
        Message_Type = Mestyp
        Creation_Date_High = Date
        Creation_Time_High = Time
    Exceptions
        Error_Code_1.
```

Example

A complex example for a function module, which collects the change pointers, can be examined in:

`MASTERIDOC_CREATE_SMD_DEBMAS` .

This one reads change pointers for debtors (customer masters). During the processing, it calls the actual IDoc creating module `MASTERIDOC_CREATE_DEBMAS` .

To summarize the change pointer concept

- Change pointers record relevant updates of transaction data
- Change pointers are written separate from the change documents, while at the same time
- Change pointers are evaluated by a collector run

BDCPS	Change pointer: Status
BDCP	Change pointer
BDCPV	A view with BDCP and BDCPS combined: Change pointer with status
TBDA2	Declare activate message types for change pointers with view V_TBDA2.or transaction *BD50* or *SALE* -> Activate change pointers for message types
TBD62	The view V_TBD62 defines those fields which are relevant for change pointer creation. The table is evaluated by the CHANGE_DOCUMENT_CLOSE function. The object is the same used by the change document. To find out the object name, look for CHANGE_DOCUMENT_CLOSE in the transaction you are inspecting or see table CDHDR for traces.

Illustration 13: **Tables involved in change pointers processing**

Sample content of view V_TBD62	Object	Table name	Field
	DEBI	KNA1	NAME3
	DEBI	Kann1	ORT01
	DEBI	Kann1	REGIO

Illustration 14: **Sample content of view V_TBD62**

9

IDoc Recipes

The chapter shall show you how an IDoc function is principally designed and how R/3 processes the IDocs. I cannot stop repeating, that writing IDoc processing routines are a pretty simple task. With a number of recipes on hand, you can easily build your own processors.

9.1 How the IDoc Engine Works

IDocs are usually created in a four step process. These steps are: retrieving the data, converting them to IDoc format, add a control record and delivering the IDoc to a port.

Collect data from R/3 database

This is the most individual task in outbound processing. You have to identify the database tables and data dependencies, which are needed in the IDoc to be sent. The smartest way is usually to select the data from database into an internal table using `SELECT * FROM dbtable INTO itab ... WHERE ...`

Wrap data in IDoc format

The collected data must be transformed into ASCII data and filled into the predefined IDoc segment structures. The segment definitions are done with transaction `WE31` and the segments allowed in an IDoc type are set up in transaction `WE30`. Segment once defined with `WE31` are automatically created as SAP DDic structures. They can be viewed with `SE11`, however they cannot be edited

Create the IDoc control record

Every IDoc must be accompanied by a control record. This record must contain at least the IDoc type to identify the syntactical structure of the data and it must contain the name and role of the sender and the receiver. This header information is checked against the partner definitions for outbound. Only if a matching partner definition exists, the IDoc can be sent. Partner definitions are set up with transaction WE20.

Send data to port

When the partner profile check passes, the IDoc is forwarded to a logical port, which is also assigned in the partner profile. This port is set up with transaction WE21 and defines the medium to transport the IDoc, e.g. file or RFC. The RFC destinations are set up with transaction SM57 and must also be entered in table TBDLS with an SM31 view. Directories for outbound locations of files are set up with transaction FILE and directly in WE21. It also allows to use a function module which generate file names. Standard functions for that purpose begin like EDI_FILE*.

9.2   How SAP Standard Processes Inbound IDocs

When you receive an IDoc the standard way, the data is stored in the IDoc base and a function module is called, which decides how to process the received information.

EDID4 - Data

Data is stored in table EDID4 (EDID3 up to release 3.xx, EDIDD up to release 2.xx)

EDIDC - Control Record

An accompanying control record with important context and administrative information is stored in table EDIDC.

Event signals readiness

After the data is stored in the IDoc base tables, an event is fired to signal that there is an IDoc waiting for processing. This event is consumed by the IDoc handler, which decides, whether to process the IDoc immediately, postpone processing or decline activity for whatever reason.

EDIFCT - Processing function

When the IDoc processor thinks it is time to process the IDoc it will have a look into table EDIFCT , where it should find the name of a function module, which will be called to process the IDoc data.

This function module is the heart of all inbound processing. The IDoc processor will call this routine and pass the IDoc data from EDID4 and the control record from EDIDC for the respective IDoc.

Function has a fixed interface

Because this routine is called dynamically it must adhere to some conventions, where the most important ones are: the interface parameters of the function must match the following call:

EDIDS - Status log

The processing steps and their respective status results are stored in table EDIDS.

Status must be logged properly

In addition the routine has to determine properly the next status of the IDoc in table EDIDS, usually it will be EDIDS-STATU = 53 for OK or 51 for error.

9.3   How To Create the IDoc Data

R/3 provides a sophisticated IDoc processing framework. This framework determines a function module, which is responsible for creating or processing the IDoc.

Function Module to generate the IDoc

The kernel of the IDoc processing is always a distinct function module. For the outbound processing the function module creates the IDoc and leaves it in an internal table, which is passed as interface parameter.

During inbound processing the function module receives the IDoc via an interface parameter table. It would interpret the IDoc data and typically update the database either directly or via a call transaction.

Function are called dynamically

The function modules are called dynamically from a standard routine. Therefore the function must adhere to a well defined interface.

Function group EDIN with useful routines

You may want to investigate the function group EDIN, which contains a number of IDoc handler routines and would call the customized function.

Copy and modify existing routines

The easiest way, to start the development of an Outbound IDoc function module, is to copy an existing one. There are many samples in the standard R/3 repository, most are named IDOC_OUTBOUND* or IDOC_OUTPUT*

Outbound sample functions are named like IDOC_OUTPUT*

```
FUNCTION IDOC_OUTPUT_ORDERS01
```

Inbound sample functions are named like IDOC_INPUT*

```
FUNCTION IDOC_INPUT_ORDERS01
```

Outbound sample functions for master data are named like MASTERIDOC_INPUT*

```
FUNCTION MASTERIDOC_CREATE_MATMAS
```

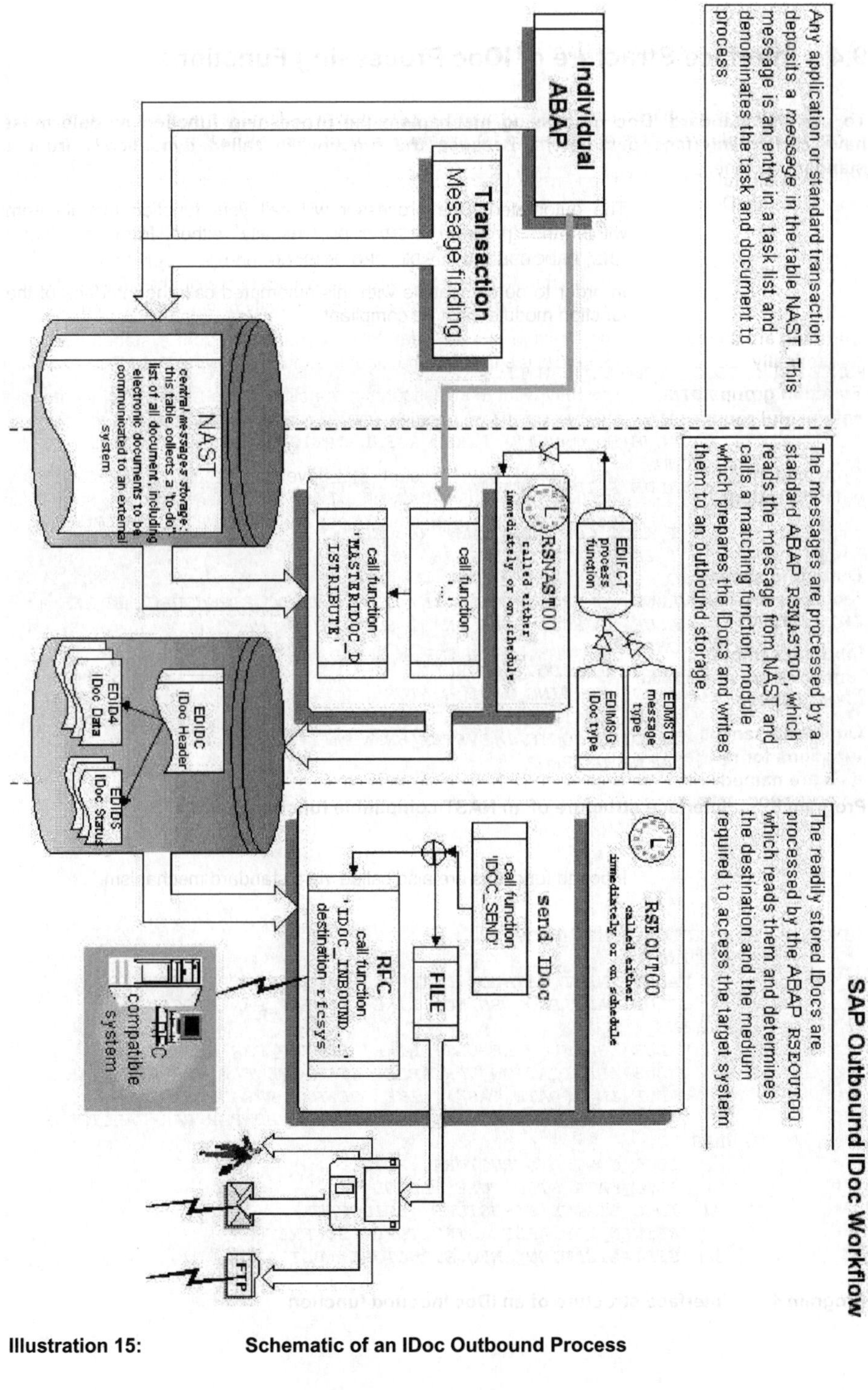

SAP Outbound IDoc Workflow

Individual ABAP

Transaction — Message finding

Any application or standard transaction deposits a *message* in the table NAST. This message is an entry in a task list and determines the task and document to process

NAST

central *messages storage*; this table collects a 'to-do' list of all document, including electronic documents to be communicated to an external system

The messages are processed by a standard ABAP RSNAST00, which reads the message from NAST and calls a matching function module which prepares the IDocs and writes them to an outbound storage.

RSNAST00
L — called either immediately or on schedule

EDI/CT process function

call function ...

call function 'MASTERIDOC_DISTRIBUTE'

EDMSG message type

EDIMSG IDoc type

EDIDC IDoc Header

EDID4 IDoc Data

EDIDS IDoc Status

The readily stored IDocs are processed by the ABAP RSEOUT00 which reads them and determines the destination and the medium required to access the target system

RSEOUT00
L — called either immediately or on schedule

send IDoc

call function 'IDOC_SEND'

FILE

RFC
call function 'IDOC_INBOUND...' destination r:csys

r:csys compatible system

FTP

Illustration 15:   **Schematic of an IDoc Outbound Process**

9.4 Interface Structure of IDoc Processing Functions

To use the standard IDoc processing mechanism the processing function module must have certain interface parameters, because the function is called dynamically from a standard routine.

The automated IDoc processor will call your function module from within the program RSNASTED, usually either from the FORM ALE_PROCESSING or EDI_PROCESSING.

In order to be compatible with this automated call, the interface of the function module must be compliant.

```
FUNCTION Z_IDOC_OUTBOUND_SAMPLE.
*"     IMPORTING
*"             VALUE(FL_TEST) LIKE  RS38L-OPTIONAL DEFAULT 'X'
*"             VALUE(FL_COMMIT) LIKE  RS38L-OPTIONAL DEFAULT SPACE
*"     EXPORTING
*"             VALUE(F_IDOC_HEADER) LIKE  EDIDC STRUCTURE  EDIDC
*"     TABLES
*"             T_IDOC_CONTRL STRUCTURE  EDIDC
*"             T_IDOC_DATA STRUCTURE  EDIDD
*"     CHANGING
*"             VALUE(CONTROL_RECORD_IN) LIKE  EDIDC STRUCTURE  EDIDC
*"             VALUE(OBJECT) LIKE  NAST STRUCTURE  NAST
*"     EXCEPTIONS
*"             ERROR_IN_IDOC_CONTROL
*"             ERROR_WRITING_IDOC_STATUS
*"             ERROR_IN_IDOC_DATA
*"             SENDING_LOGICAL_SYSTEM_UNKNOWN
*"             UNKNOWN_ERROR
```

Program 3:   Interface structure of an NAST compatible function module

Inbound functions are also called via a standard mechanism.

```
FUNCTION IDOC_INPUT_SOMETHING.
*"     IMPORTING
*"             VALUE(INPUT_METHOD) LIKE  BDWFAP_PAR-INPUTMETHD
*"             VALUE(MASS_PROCESSING) LIKE  BDWFAP_PAR-MASS_PROC
*"     EXPORTING
*"             VALUE(WORKFLOW_RESULT) LIKE  BDWFAP_PAR-RESULT
*"             VALUE(APPLICATION_VARIABLE) LIKE  BDWFAP_PAR-APPL_VAR
*"             VALUE(IN_UPDATE_TASK) LIKE  BDWFAP_PAR-UPDATETASK
*"             VALUE(CALL_TRANSACTION_DONE) LIKE  BDWFAP_PAR-CALLTRANS
*"     TABLES
*"             IDOC_CONTRL STRUCTURE  EDIDC
*"             IDOC_DATA STRUCTURE  EDIDD
*"             IDOC_STATUS STRUCTURE  BDIDOCSTAT
*"             RETURN_VARIABLES STRUCTURE  BDWFRETVAR
*"             SERIALIZATION_INFO STRUCTURE  BDI_SER
```

Program 4:   Interface structure of an IDoc inbound function

9.5   Recipe To Develop An Outbound IDoc Function

This is an individual coding part where you need to retrieve the information from the database and prepare it in the form the recipient of the IDoc will expect the data

Read data to send   The first step is reading the data from the database, the one you want to send.

```
FUNCTION Y_AXX_COOKBOOK_TEXT_IDOC_OUTB.
*"----------------------------------------------------------------------
*"*"Lokale Schnittstelle:
*"      IMPORTING
*"             VALUE(I_TDOBJECT) LIKE  THEAD-TDOBJECT DEFAULT 'TEXT'
*"             VALUE(I_TDID) LIKE  THEAD-TDID DEFAULT 'ST'
*"             VALUE(I_TDNAME) LIKE  THEAD-TDNAME
*"             VALUE(I_TDSPRAS) LIKE  THEAD-TDSPRAS DEFAULT SY-LANGU
*"      EXPORTING
*"             VALUE(E_THEAD) LIKE  THEAD STRUCTURE  THEAD
*"      TABLES
*"             IDOC_DATA STRUCTURE  EDIDD OPTIONAL
*"             IDOC_CONTRL STRUCTURE  EDIDC OPTIONAL
*"             TLINES STRUCTURE  TLINE OPTIONAL
*"      EXCEPTIONS
*"             FUNCTION_NOT_EXIST
*"             VERSION_NOT_FOUND
*"----------------------------------------------------------------------
  CALL FUNCTION 'READ_TEXT'
       EXPORTING
             ID                     = ID
             LANGUAGE               = LANGUAGE
             NAME                   = NAME
             OBJECT                 = OBJECT
       TABLES
             LINES                  = LINES.
* now stuff the data into the IDoc record format
  PERFORM PACK_LINE TABLES IDOC_DATA USING 'THEAD' E_THEAD.
  LOOP AT LINES.
    PERFORM PACK_LINE TABLES IDOC_DATA USING 'THEAD' LINES.
  ENDLOOP.
ENDFUNCTION.
```

9.6   Converting Data Into IDoc Segment Format

The physical format of the IDocs records is always the same. Therefore the application data must be converted into a 1000 character string.

Fill the data segments which make up the IDoc
An IDocs is a file with a rigid formal structure. This allows the correspondents to correctly interpret the IDoc information. Were it for data exchange between SAP-systems only, the IDoc segments could be simply structured like the correspondent DDIC structure of the tables whose data is sent.

However, IDocs are usually transported to a variety of legacy systems which do not run SAP. Both correspondents therefore would agree an IDoc structure which is known to the sending and the receiving processes.

Transfer the whole IDoc to an internal table, having the structure of EDIDD
All data needs to be compiled in an internal table with the structure of the standard SAP table EDIDD. The records for EDIDD are principally made up of a header string describing the segment and a variable length character field (called SDATA) which will contain the actual segment data.

```
FORM PACK_LINE TABLES IDOC_DATA USING 'THEAD' E_THEAD.
  TABLES: THEAD.
  MOVE-CORRESPONDING E:THEAD to Z1THEAD.
  MOVE ,Z1THEAD' TO IDOC_DATA-SEGNAM.
  MOVE Z1THEAD TO IDOC_DATA-SDATA.
  APPEND IDOC_DATA.
ENDFORM. "
```

Program 5:   Routine to move the translate to IDoc data

Fill control record
Finally the control record has to be filled with meaningful data, especially telling the IDoc type and message type.

```
  IF   IDOC_CONTRL-SNDPRN IS INITIAL.
    SELECT SINGLE * FROM T000 WHERE MANDT EQ SY-MANDT.
    MOVE T000-LOGSYS TO IDOC_CONTRL-SNDPRN.
  ENDIF.
  IDOC_CONTRL-SNDPRT = 'LS'.
* Trans we20 -> Outbound Controls muss entsprechend gesetzt werden.
* 2  = Transfer IDoc immediately
* 4  = Collect IDocs
  IDOC_CONTRL-OUTMOD = '2'.       "1=imediately, subsystem
  CLEAR IDOC_CONTRL.
  IDOC_CONTRL-IDOCTP = 'YAXX_TEXT'.
  APPEND IDOC_CONTRL.
```

Program 6:   Fill the essential information of an IDoc control record

| 10 |

Partner Profiles and Ports

R/3 defines partner profiles for every EDI partner. The profiles are used to declare the communication channels, schedule and conditions of processing.

Summary

- Partner profiles declare the communication medium to be used with a partner
- Ports define the physical characteristics of a communication channel
- If you define an ALE scenario for your IDoc partners, you can use the ALE automated partner profile generation (→ ALE)

10.1 IDoc Type and Message Type

An IDoc file requires a minimum of accompanying information to give sense to it. These are the message type and the IDoc type. While the IDoc type tells you about the fields and segments of the IDoc file, the message type flags the context under which the IDoc was sent.

IDoc Type signals Syntactical Structure

A receiver of an IDoc must exactly know the syntactical structure of the data package received. Naturally, the receiver only sees a text file with lines of characters. In order to interpret it, it is necessary to know, which segment types the file may content and how a segment is structured into fields. SAP sends the name of the IDoc type in the communication header.

IDoc type (WE30)

The IDoc type describes the file structure. The IDoc type is defined and viewable with transaction `WE30`.

Examples:

- Examples of IDoc types are: `MATMAS01`, `ORDERS01`, `COND_A01` or `CLSMAS01`.

Message Type signal the semantic context

The message type is an identifier that tags the IDoc to tell the receiver, how the IDoc is meant to be interpreted. It is therefore the tag for the semantic content of the IDoc.

Examples

- Examples of IDoc types are: `MATMAS`, `ORDERS`, `COND_A` or `CLSMAS`.

For any combination of message type and receiving partner, a profile is maintained

The combination of IDoc type and message type gives the IDoc the full meaning. Theoretically you could define only a single IDoc type for every IDoc you send. Then, all IDocs would have the same segments and the segments would have always the same field structure. According to the context some of the record fields are filled, others are simply void. Many ancient interfaces are still working that way.

Typical combinations of IDoc and message types are the following:

	Message Type	IDoc Type
Sales order, older format	ORDERS	ORDERS01
Sales order, newer format	ORDERS	ORDERS02
Purchase Requisition	PURREQ	ORDERS01

The example shows you, that sales orders can be exchanged in different file formats. There may be some customers who accept the latest IDoc format `ORDERS02`, while others still insist in receiving the old format `ORDERS01`.

The IDoc format for sales orders would also be used to transfer a purchase requisition. While the format remains the same, the different message type signals, that this is not an actual order but a request.

10.2 Partner Profiles

Partner profiles play an important role in EDI communications. They are parameter files which store the EDI partner dependent information.

Partner profile define the type of data and communication paths of data to be exchanged between partner	When data is exchanged between partners it is important that sender and receiver agree about the exact syntax and semantics of the data to be exchanged. This agreement is called a *partner profile* and tells the receiver the structure of the sent file and how its content is to be interpreted.
	The information defined with the partner profile are:
For any combination of message type and receiving partner, a profile is maintained	IDoc type and message type as key identifier of the partner profile
	Names of sender and receiver to exchange the IDoc information for the respective IDoc and message type and
	Logical port name via which the sender and receiver, resp. will communicate
The communication media is assigned by the profile	If you exchange e.g. sales orders with partners, you may do this via different media with different customers. There may be one customer to communicate with you via TCP/IP (the Internet) while the other still insists in receiving diskette files.
Profiles cannot be transported	They must be defined for every R/3 client individually. They cannot be transported using the R/3 transport management system. This is because the profile contain the name of the sending system, which are naturally different for consolidation and production systems.
Profiles define the allowed EDI connections	The profiles allow you to open and close EDI connection with individual partners and specify in detail which IDocs are to be exchanged via the interface.
Profiles can also used to block an EDI communication	The profile is also the place to lock permanently or temporarily an IDoc communication with an EDI partner. So you shut the gate for external communication with the profile.

http://idocs.de http://logosworld.de

10.3 Defining the partner profile (`WE20`)

The transaction WE20 is used to set up the partner profile.

`WE20`

The profiles are defined with transaction `WE20`, which is also found in the EDI master menu `WEDI`. From there you need to specify partner and partner type and whether you define a profile for inbound or outbound. Additionally you may assign the profile to a NAST message type.

**Partner type, e.g.
LI=Supplier
CU=Customer
LS=Logical system**

The partner type defines from which master data set, the partner number originates. The partner types are the ones which are used in the standard applications for SD, MM or FI. The most important types for EDI are LI (=Lieferant, supplier), CU (Customer) or LS (Logical system). The logical system is of special interest, when you exchange data with computer subsystems via ALE or other RFC means.

Inbound and Outbound definitions

For every partner and every direction of communication, whether you receive or send IDocs, a different profile is maintained. The inbound profile defines the processing routine. The outbound profile defines mainly the target, where to send the data to.

Link message type to outbound profile

If you send IDocs out of an application's messaging, i.e. a communication via the NAST table, then you have to link the message type with an IDoc profile. This is also done in transaction WE20..

Inbound profiles determine the processing logic

The processing code is a logical name for the processing function module or object method. The processing code is used to uniquely determine a function module that will process the received IDoc data. The inbound profile will point to a processing code.

10.4 Data Ports (*WE21*)

IDoc data can be sent and received through a multitude of different media. In order to decouple the definition of the media characteristics from the application using it, the media is accessed via ports.

A port is a logical name to access a physical input/output device

A port is a logical name for an input/output device. A program talks to a port which is presented to it with a common standard interface. The port takes care of the translation between the standard interface format and the device dependent format.

Communication media is defined via a port definition

Instead of defining the communication path directly in the partner profile, a port number is assigned rather. The port number then designates the actual medium. This allows to define the characteristics of a port individually and use that port in multiple profiles. Changes in the port will than reflect automatically to all profiles without touching them.

Typical ports for data exchange are:

Communication Media
- Disk file with a fixed name
- Disk file with dynamic names
- Disk file with trigger of a batch routine
- Standard RFC connection via TCP/IP
- A network channel
- TCP/IP FTP destination (The Internet)
- Call to a individual program e.g. EDI converter

Every program should communicate with other computers via the ports only

Every application should send or receive its data via the logical ports only. This allows to easily change the hardware and software used to make the physical I/O connection without interfering with the programs itself.

The transactions used to define the ports are

WE21 defines the port; SM59 sets up media

| *WE21* | to create the port and assign a logical name |
| *SM59* | to define the physical characteristics of the I/O device used |

There are different port versions for the respective R/3 releases as shown in the matrix below:

Port Type	DDic Format	Release
1	not used	not used
2	EDID3	2.x, 3.x
3	EDID4	4.x

Illustration 16:   **R/3 port types by release**

Port versions differ in length of fields

The difference between the port types are mainly the length of some fields. E.g. does port type 3 allow segment names up to 30 characters in length, while port type 3 is constraint to a maximum segment name of 8 characters.

11

Workflow Technology

There are two faces of workflow in R/3. There is once the business oriented workflow design as it is taught in universities. This is implemented by the SAP Business Workflow™. However, the workflow is also a tool to link transactions easily. It can be used to define execution chains of transactions or to trigger user actions without the need to modify the SAP standard code. This does not require to laboriously customize the HR related workflow settings.

Summary

- Workflow event linkage allows the execution of another program when a transaction finishes

- The workflow event linkage mechanism can be easily used without customizing the full workflow scenarios

- This way we use the workflow engine to chain the execution of transaction and circumvent the set-up of the *SAP Business Workflow™*

- There are several independent ways to trigger the workflow event linkage

Americans work hard because they are optimists.
Germans work hard because they fear the future.

11.1 Workflow in R/3 and Its Use For Development

SAP R/3 provides a mechanism, called Workflow, that allows conditional and unconditional triggering of subsequent transactions from another transaction. This allows to build up automatic processing sequences without having the need to modify the SAP standard transactions.

Workflow is a business planning method

The SAP business workflow was originally designed to model business workflows according to scientific theories called **Business Workflow**. Business workflows are mainly a modelling tool, that uses graphical means, e.g. flow charting, to sketch the flow of events in a system. SAP allows to transcript these event modelling into customizing entries, which are then executed by the SAP Workflow mechanism.

Transaction SWO1

The transaction to enter the graphical model, to define the events and objects and to develop necessary triggering and processing objects, is SWO1 (it is an O not a zero).

SAP approach unnecessary complex and formal

I believe, that the way how workflows are realized in SAP is far to complicated and unnecessarily complex and will fill a separate book.

Workflow events can be used for own developments

Fortunately the underlying mechanism for workflows is less complex as the formal overhead. Most major transactions will trigger the workflow via `SWE_EVENT_CREATE`. This will make a call to a workflow handler routine, whose name can usually be customized dynamically and implemented as a function module.

11.2 Event Coupling (Event Linkage)

Contrary to what you mostly hear about R/3 workflow, it is relatively easy and mechanical to define a function module as a consecutive action after another routine raised a workflow event. This can e.g. be used to call the execution of a transaction after another one has finished.

Every workflow enabled transaction will call SWE_EVENT_CREATE

The whole workflow mechanism is based on a very simple principle. Every workflow enabled transaction will call directly or indirectly the function module SWE_EVENT_CREATE during update.

SWE_EVENT_CREATE will look in a table, e.g. SWETYPECOU to get the name of the following action

The function module SWE_EVENT_CREATE will then consult a customizing table. For a simple workflow coupling, the information is found in the table SWETYPECOU . The table will tell the name of the subsequent program to call, either a function module or an object method.

This way of defining the subsequent action is called type coupling because the action depends on the object type of the calling event.

The call to the following event is done with a dynamic function call. This requires, that the called function module has a well-defined interface definition. The examples shows the call as it is found in SWE_EVENT_CREATE .

```
CALL FUNCTION typecou-recgetfb " call receiver_type_get_fb
     EXPORTING
          objtype = typecou-objtype
          objkey = objkey
          event = event
          generic_rectype = typecou-rectype
     IMPORTING
          rectype = typecou-rectype
     TABLES
          event_container = event_container
     EXCEPTIONS
     OTHERS = 1.
```

Program 7:   This is the call of the type coupled event in release 40B

11.3 Workflow from Change Documents

Every time a change document is written a workflow event for the change document object is triggered. This can be used to chain unconditionally an action from a transaction.

CHANGEDOCUMENT_CLOSE The most interesting chaining point for workflow events is the creation of the change document. Nearly every transaction writes change documents to the database. This document is committed to the database with the function module CHANGEDOCUMENT_CLOSE. This function will also trigger a workflow event.

The workflow handler triggered by an event which is fired from change documents is defined in table SWECDOBJ. For every change document type a different event handler can be assigned. This is usually a function module and the call for it is the following

```
CALL FUNCTION swecdobj-objtypefb
    EXPORTING
        changedocument_header = changedocument_header
        objecttype = swecdobj-objtype
    IMPORTING
        objecttype = swecdobj-objtype
    TABLES
        changedocument_position = changedocument_position.
```

Program 8:   This is the call of the change doc event in release 40B

In addition change pointers for ALE are written

Change pointers are created by calling FUNCTION CHANGEDOCUMENT_CLOSE, which writes the usual change documents into table CDHDR and CDPOS. This function calls then the routine CHANGE_POINTERS_CREATE which create the change pointers.

```
CALL FUNCTION 'CHANGE_POINTERS_CREATE'
    EXPORTING
        change_document_header = cdhdr
    TABLES
        change_document_position = ins_cdpos.
```

Program 9:   This is the call of the type coupled event in release 40B

11.4 Trigger a Workflow from Messaging

The third common way to trigger a workflow is doing it from messaging.

Define a message for condition technique

When the R/3 messaging creates a message and processes it immediately, then it actually triggers a workflow. You can use this to set up conditional workflow triggers, by defining a message with the message finding and link the message to a workflow.

Assign media W or 8

You define the message the usual way for your application as you would do it for defining a message for SAPscript etc. As a processing media you can assign type W for workflow or 8 for special processing.

The media type W for workflow would require defining an object in the object repository. We will only show how you can trigger the workflow with a standard ABAP using the media type 8.

Form routine requires two parameters

You need to assign a program and a form routine to the message in table TNAPR. The form routine you specify needs exactly two USING-parameters as in the example below.

```
REPORT ZSNASTWF.
TABLES: NAST.

FORM ENTRY USING RETURN_CODE US_SCREEN.
*      Here you gonna call your workflow action
  RETURN_CODE = 0.
  SY-MSGID = '38'.
  SY-MSGNO = '000'.
  SY-MSGNO = 'I'.
  SY-MSGV1 = 'Workflow called via NAST'.
  CALL FUNCTION 'NAST_PROTOCOL_UPDATE'
       EXPORTING
           MSG_ARBGB = SYST-MSGID
           MSG_NR    = SYST-MSGNO
           MSG_TY    = SYST-MSGTY
           MSG_V1    = SYST-MSGV1
           MSG_V2    = SYST-MSGV2
           MSG_V3    = SYST-MSGV3
           MSG_V4    = SYST-MSGV4
       EXCEPTIONS
           OTHERS    = 1.
*   ---> Insert your program code here
ENDFORM.
```

Illustration 17:   **Fully functional NAST event handler**

NAST must be declared public in the called program

In addition, you need to declare the table NAST with a tables statement public in the ABAP where the form routine resides. When the form is called the variable NAST is filled with the values of the calling NAST message. The program shown here is a fully functional NAST handler. What you need to do is to add your own processing logic.

Call transaction can only performed on commit or in a separate task

If you intend to do a call transaction in the handler, than you have to take into account, that the NAST handler can be called in an update task. This would cause the call transaction to crash. To avoid this, you should put the call transaction in a separate subroutine and call it on commit, e.g. PERFORM call_trans ON COMMIT. Of course you could call a function module in background task as well.

11.5 Example, How To Create A Sample Workflow Handler

Let us show you a function module which is suitable to serve as a function module and define the linkage.

Create a function module that will be triggered by a workflow event

We want to create a very simple function module that will be triggered by a workflow event. This function is called from within function SWE_EVENT_CREATE. The parameters must comply with the calling standard as shown below.

This is the call of the type coupled event in release 40B

```
CALL FUNCTION typecou-recgetfb
     EXPORTING
             objtype = typecou-objtype
             objkey = objkey
             event = event
             generic_rectype = typecou-rectype
     IMPORTING
             rectype = typecou-rectype
     TABLES
             event_container = event_container
     EXCEPTIONS
     OTHERS = 1.
```

Template for workflow handler

Release   40B   provides   the   function   module WF_EQUI_CHANGE_AFTER_ASSET which could be used as template for the interface. So we will copy it and put our coding in instead..

```
FUNCTION Z_WORKFLOW_HANDLER.
*"*"Lokale Schnittstelle:
*"       IMPORTING
*"             VALUE(OBJKEY) LIKE  SWEINSTCOU-OBJKEY
*"             VALUE(EVENT) LIKE  SWETYPECOU-EVENT
*"             VALUE(RECTYPE) LIKE  SWETYPECOU-RECTYPE
*"             VALUE(OBJTYPE) LIKE  SWETYPECOU-OBJTYPE
*"       TABLES
*"             EVENT_CONTAINER STRUCTURE  SWCONT
*"       EXCEPTIONS
*"             NO_WORKFLOW
  RECEIVERS-EXPRESS  = ' '.
  RECEIVERS-RECEIVER = SY-SUBRC.
  APPEND RECEIVERS.
  DOCUMENT_DATA-OBJ_DESCR
= OBJ_KEY.
  CONTENT = OBJ_KEY.
  APPEND CONTENT.
  CALL FUNCTION 'SO_NEW_DOCUMENT_SEND_API1'
       EXPORTING  DOCUMENT_DATA          = DOCUMENT_DATA
       TABLES     OBJECT_CONTENT         = CONTENT
                  RECEIVERS              = RECEIVERS.
ENDFUNCTION.
```

Program 10:   A workflow handler that sends an Sap Office mail

Link handler to caller

The function can be registered as a handler for an event. This is done with transaction SWLD.

Event logging

If you do not know the object type, that will trigger the event, you can use the event log to find out. You have to activate the log from $SWLD$ and then execute the event firing transaction. When the event has been fired it will a write a trace in the event log.

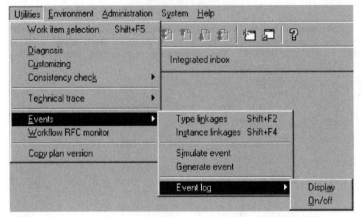

Illustration 14: Transaction SWLD to define event linkage and see event log

11.6 Troubleshooting Workflow Events

All workflow handlers are called via RFC to a dummy destination WORKFLOW_LOCAL_000 where 000 is to be replaced by the client number.

Most errors are caused by following reasons

Hit list of common errors

- You forgot to set the RFC flag in the interface definition of your event handling function module
- There is a syntax error in your function module (check with generate function group)
- You mistyped something when defining the coupling
- The internal workflow destination WORKFLOW_LOCAL_000 is not defined

SM58 to display what happened to your event

If you think your handler did not execute at all, you can check the list of pending background tasks with transaction SM58. If you event is not there it has either neither been triggered (so your tables SWETYPEENA and SSWETYPEOBJ may have the wrong entries) or your event handler executed indeed and may probably have done something else than you expected. Ergo: your mistake.

Read carefully the help for CALL FUNCTION .. IN BACKGROUND TASK

Your event handler function is called IN BACKGROUND TASK. You may want to read carefully the help on this topic in the SAP help. (help for „Call function" from the editor command line)

11.7   Send A SAP Mail From A Workflow Event

This is an example coding to demonstrate how you can send a SAP mail triggered by a workflow event.

```
FUNCTION YAXXWF_MAIL_ON_EVENT.
*"       IMPORTING
*"             VALUE(OBJKEY) LIKE   SWEINSTCOU-OBJKEY
*"             VALUE(EVENT) LIKE   SWETYPECOU-EVENT
*"             VALUE(RECTYPE) LIKE   SWETYPECOU-RECTYPE
*"             VALUE(OBJTYPE) LIKE   SWETYPECOU-OBJTYPE
*"       TABLES
*"             EVENT_CONTAINER STRUCTURE   SWCONT
*-------------------------------------------------------------*
* This example sends a mail to the calling user and tells
* about the circumstances when the event was fired.
* Just for fun, it lists also all current enqueue locks
*-------------------------------------------------------------*
  DATA: ENQ    LIKE SEQG3  OCCURS 0 WITH HEADER LINE.
  DATA: DOC_DATA LIKE SODOCCHGI1.
  DATA: MAIL LIKE STANDARD TABLE OF SOLISTI1 WITH HEADER LINE.
  DATA: RECLIST LIKE STANDARD TABLE OF SOMLRECI1 WITH HEADER LINE.
  MAIL-LINE    = ,Event fired by user: &'.
  REPLACE ,&' WITH SY-UNAME INTO MAIL-LINE.
  APPEND MAIL.
*-------------------------------------------------------------*
  MAIL-LINE    = ,Object Key: &'.
  REPLACE ,&' WITH OBJKEY INTO MAIL-LINE.
  APPEND MAIL.
*-------------------------------------------------------------*
  MAIL-LINE    = ,Event Name: &'.
  REPLACE ,&' WITH EVENT  INTO MAIL-LINE.
  APPEND MAIL.
*-------------------------------------------------------------*
  MAIL-LINE    = ,Rectype: &'.
  REPLACE ,&' WITH RECTYPE INTO MAIL-LINE.
  APPEND MAIL.
*-------------------------------------------------------------*
  MAIL-LINE    = ,Object Type: &'.
  REPLACE ,&' WITH OBJTYPE INTO MAIL-LINE.
  APPEND MAIL.
*-------------------------------------------------------------*
  MAIL-LINE    = ,Container contents:'.
  APPEND MAIL.
*-------------------------------------------------------------*
  LOOP AT EVENT_CONTAINER.
    CONCATENATE EVENT_CONTAINER-ELEMENT EVENT_CONTAINER-VALUE
          INTO MAIL-LINE SEPARATED BY SPACE.
    APPEND MAIL.
  ENDLOOP.
*----- write the current enqueues into the message -(for demo)---------
  MAIL-LINE    = ,Active enqueue locks when event was triggered:'.
  APPEND MAIL.
  CALL FUNCTION ,ENQUEUE_READ' TABLES ENQ = ENQ.
  LOOP AT ENQ.
    CONCATENATE ENQ-GNAME ENQ-GARG    ENQ-GMODE  ENQ-GUSR  ENQ-GUSRVB
              ENQ-GOBJ  ENQ-GCLIENT ENQ-GUNAME ENQ-GTARG ENQ-GTCODE
    INTO MAIL-LINE SEPARATED BY ,/'.
```

http://idocs.de http://logosworld.de

```
   APPEND MAIL.
 ENDLOOP.
 IF ENQ[] IS INITIAL.
   MAIL-LINE = ,*** NONE ***'.
   APPEND MAIL.
 ENDIF.
*----------------------------------------------------------------------*
* fill the receiver list
 REFRESH RECLIST.
 RECLIST-RECEIVER = ,USERXYZ'.
 RECLIST-REC_TYPE = ,B'.
 RECLIST-EXPRESS  = , ,.
 reclist-express = ,X'. "will pop up a notification on receiver screen
 APPEND RECLIST.
*----------------------------------------------------------------------*
 CLEAR DOC_DATA.
 DOC_DATA-OBJ_NAME   = ,WF-EVENT'.
 DOC_DATA-OBJ_DESCR  = ,Event triggered by workflow type coupling'.
 DOC_DATA-OBJ_SORT   = ,WORKFLOW'.
 doc_data-obj_expdat
 doc_data-sensitivty
 doc_data-obj_prio
 doc_data-no_change
*----------------------------------------------------------------------*
 CALL FUNCTION ,SO_NEW_DOCUMENT_SEND_API1'
      EXPORTING
           DOCUMENT_DATA                 = DOC_DATA
*          DOCUMENT_TYPE                 = ,RAW'
*          PUT_IN_OUTBOX                 = , ,
      IMPORTING
*          SENT_TO_ALL                   =
*          NEW_OBJECT_ID                 =
      TABLES
*          OBJECT_HEADER                 =
           OBJECT_CONTENT                = MAIL
*          OBJECT_PARA                   =
*          OBJECT_PARB                   =
           RECEIVERS                     = RECLIST
      EXCEPTIONS
           TOO_MANY_RECEIVERS            = 1
           DOCUMENT_NOT_SENT             = 2
           DOCUMENT_TYPE_NOT_EXIST       = 3
           OPERATION_NO_AUTHORIZATION    = 4
           PARAMETER_ERROR               = 5
           X_ERROR                       = 6
           ENQUEUE_ERROR                 = 7
           OTHERS                        = 8.
*----------------------------------------------------------------------*
ENDFUNCTION.
```

Program 11:   Send a SAPoffice mail triggered by a workflow event (full example)

Calling R/3 Via OLE/JavaScript

Using the OLE/Active-X functionality of R/3 you can call R/3 from any object aware language. Actually it must be able to do DLL calls to the RFC libraries of R/3. SAP R/3 scatters the documentation for these facilities in several subdirectories of the SAPGUI installation. For details you have to look for the SAPGUI Automation Server and the SDK (RFC software development kit).

Summary

- R/3 can exchange its IDoc by calling a program that resides on the server
- The programs can be written in any language that supports OLE-2/Active-X technology
- Programming skills are mainly required on the PC side, e.g. you need to know Delphi, JavaScript or Visual Basic well

12.1 R/3 RFC from MS Office Via Visual Basic

The Microsoft Office suite incorporates with Visual Basic for Applications (VBA) a fully object oriented language. JavaScript and JAVA are naturally object oriented. Therefore you can easily connect from JavaScript, JAVA, WORD, EXCEL and all the other VBA compliant software to R/3 via the CORBA compatible object library (in WINDOWS known also DLLs or ACTIVE-X (=OLE/2) components).

Visual Basic is DCOM compliant

Visual Basic is finally designed as an object oriented language compliant to DCOM standard.

JavaScript or JAVA are object languages

JavaScript is a typical object oriented language which is compliant to basic CORBA, DCOM and other popular object standards.

SAP R/3 provides a set of object libraries, which can be registered with Visual Basic. The library adds object types to VBA which allow RFC calls to R/3.

DLLs installed with SAPGUI

The libraries are installed to the workstation with the SAPGUI installation. They are technically public linkable objects, in WINDOWS these are DLLs or ACTIVE-X controls (which are DLLs themselves).

Object library SAP provides a method CALL which will call a function module with all interface parameters

The object library SAP contains among others the object type FUNCTIONS whose basic method CALL performs an RFC call to a specified R/3 function module. With the call you can pass object properties which will be interpreted as the interface parameters of the called function module.

If the RFC call appear not to be working, you should first try out to call one of the standard R/3 RFC function like RFC_CALL_TRANSACTION_USING (calls a specified transaction or RFC_GET_TABLE (returns the content of a specified R/3 database table).

SAP R/3 provides a set of object libraries, which can be registered with JavaScript to allow RFC calls to R/3.

The object library SAP contains among others the object type FUNCTIONS whose basic method CALL performs an RFC call to a specified R/3 function module.

Try to call standard routines for testing

If the RFC call appears to be not working, you should first try out to call one of the standard R/3 RFC functions like RFC_CALL_TRANSACTION_USING (calls a specified transaction) or RFC_GET_TABLE (returns the content of a specified R/3 database table).

12.2 Call Transaction From Visual Basic for WORD 97

This is a little WORD 97 macro, that demonstrates how R/3 can be called with a mouse click directly from within WORD 97.

The shown macro calls the function module `RFC_CALL_TRANSACTIION_USING` . This function executes a dynamic call transaction using the transaction code specified as the parameter.

You can call the macro from within word, by attaching it to a pseudo-hyperlink. This is done by adding a MACROBUTTON field to the WORD text. The `macrobutton` statement must call the VBA macro R3CallTransaction and have as the one and only parameter the name of the requested transaction

`MACROBUTTON R3CallTransaction VA02`

This will call transaction VA02 when you click on the `macrobutton` in the text document. You can replace VA02 with the code of your transaction.

For more information see the Microsoft Office help for `MACROBUTTON` and Visual Basic.

Calling SAP R/3 from within WORD 97 with a mouse click

Word 97 Macro by Axel Angeli Logos! Informatik GmbH D-68782 Bruehl
From website http://www.logosworld.com

This WORD 97 document contains a Visual Basic Project which allows to call SAP R/3 transaction using the SAP automation GUI. The call is done via the WORD field insertion MACROBUTTON. You must have the SAP Automation GUI or SAP RFC Development Kit installed on your workstation to give SAP the required OLE functionality.

Example:
Click to start transaction { MACROBUTTON R3CallTransaction VA02 }
and another call to { MACROBUTTON R3CallTransaction VA02 } .

To show the coding of the MACROBUTTON statement, right-mouse-click on the transaction code link and choose "Toggle Field Codes".

Illustration 18:   **WORD 97 text with MACROBUTTON field inserted**

```
Dim fns As Object
Dim conn As Object
Dim SAP_logon As Boolean
Sub R3CallTransaction()
' get the TCODE from the WORD text, MACROBUTTON does not allow parameters
  tcode = Selection.Text & ActiveDocument.Fields(1).Code
  ll = Len("MACROBUTTON R3CallTransaction ") + 3
  tcode = Mid$(tcode, ll)
  R3CallTransactionExecute (tcode)
End Sub
Sub R3CallTransactionExecute(tcode)
On Error GoTo ErrCallTransaction
  R3Logon_If_Necessary
  Result = fns.RFC_CALL_TRANSACTION(Exception, tcode:=tcode)
  the_exception = Exception
  ErrCallTransaction: ' Error Handler General
      Debug.Print Err
      If Err = 438 Then
          MsgBox "Function module not found or RFC disabled"
          R3Logoff ' Logoff to release the connection !!!
          Exit Sub
      Else
          MsgBox Err.Description
      End If
End Sub
Sub R3Logon_If_Necessary()
  If SAP_logon <> 1 Then R3Logon
End Sub
Sub R3Logon()
      SAP_logon = False
      Set fns = CreateObject("SAP.Functions") ' Create functions object
      fns.logfilename = "wdtflog.txt"
      fns.loglevel = 1
      Set conn = fns.connection
      conn.ApplicationServer = "r3"
      conn.System = "DEV"
      conn.user = "userid"
      conn.Client = "001"
      conn.Language = "E"
      conn.tracelevel = 6
      conn.RFCWithDialog = True
      If conn.logon(0, False) <> True Then
          MsgBox "Cannot logon!."
          Exit Sub
      Else
        SAP_logon = conn.IsConnected
      End If
End Sub
Sub R3Logoff()
    conn.logoff
    SAP_logon = False
End Sub
```

Illustration 19: **Visual Basic code with macros to call R/3 from WORD 97**

12.3 R/3 RFC from JavaScript

JavaScript is a fully object oriented language. Therefore you can easily connect from JavaScript to R/3 via the CORBA compatible object library (in WINDOWS known also DLLs or ACTIVE-X (=OLE/2) components).

JavaScript is a typical object oriented language which is compliant to basic CORBA, DCOM and other popular object standards.

SAP R/3 provides a set of object libraries, which can be registered with JavaScript to allow RFC calls to R/3.

DLLs installed with SAPGUI

The libraries are installed to the workstation with the SAPGUI installation.

The object library SAP contains among others the object type FUNCTIONS whose basic method CALL performs an RFC call to a specified R/3 function module.

Try to call standard routines for testing

If the RFC call appears to be not working, you should first try out to call one of the standard R/3 RFC functions like RFC_CALL_TRANSACTION_USING (calls a specified transaction) or RFC_GET_TABLE (returns the content of a specified R/3 database table).

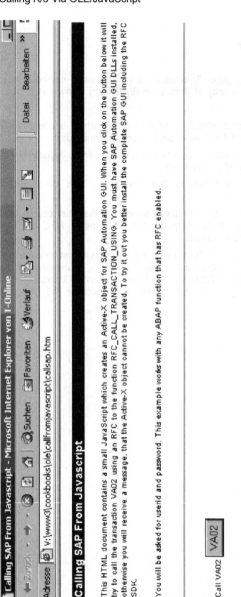

Illustration 15:   **HTML Page with a button to call a transaction via RFC**

```
<script language="JavaScript">
<!--
retcd = 0;
exceptions = 0;
// ***      SAPLogon() creates an object that has the methods to
//     execute a call to an SAP function module
function SAPlogon()
    { fns                = new ActiveXObject("SAP.Functions");
      trans              = fns.Transactions;
      conn               = fns.connection;    /* get a new connection
object */
      conn.System        = "DEV";      /* Set the system ID (see: SY-SYSID)
*/
      conn.user   = "userid"; /* set userid (blank for dialog) */
      conn.password      = ""; /* set password (blank for dialog) */
      conn.Client        = "100";      /* set password (blank for dialog) */
      conn.Language      = "E";        /* set language (blank for default)
*/
      conn.tracelevel    = 6;    /* set password (blank for dialog) */
      conn.RFCWithDialog = 1;    /* true: opens visible session window */
      exceptions = 0;
      conn.logon(0, 0);          /* *** this call creates the object *** */
    };
function SAPlogoff()
    { conn.logoff(0, 0);
      exceptions = 0;
    };
// ***  execute the SAP function MODULE "RFC_CALL_TRANSACTION_USING"
//     as a method execution of object type SAP.functions
function SAPcallTransaction(tcode)
    { exceptions              = 0;
      callta                  = fns.add("RFC_CALL_TRANSACTION_USING");
      callta.exports("TCODE") = "VA02";
      callta.exports("MODE")  = "E";
      retcd                   = callta.call;
      conn.logoff();
      alert(retcd);
      SAPcallTransaction      = retcd;
    };
// --></script>
<body>
<!—Create an HTML button with a JavaScript call attached -->
Call VA02
<input TYPE   = "submit"
       VALUE  = "VA02"
       OnClick = "SAPlogon();
                 SAPcallTransaction(&quot;VA02&quot;);
                 SAPlogoff()"
>
</body>
```

Program 12:   JavaScript example to call an R/3 function module via OLE/RFC

12.4 R/3 RFC from Visual Basic Script

Visual Basic Script also known as Windows Scripting Host is the scripting tool which is delivered with any Windows installation. See here an extensive example to read any table from R/3.

```
'****************************************************************
' (c) 2000, Axel Angeli
' The R/3 Guide To ecommerce with mySAP.com, Visual Basic and WWW
'----------------------------------------------------------------
' This is a little VB Script example that demonstrates in a very simple
way
' how you can call a RFC function module in R/3. The function used is
' RFC_READ_TABLE which takes the name of a table as a parameter and
' returns its contents. As on option you can pass a Tablefactory element
' to contain a simple SQL WHERE-clause, which is added to the SQL
statement issued by
' RFC_READ_TABLE
'----------------------------------------------------------------
' If you fully understood, how this example works, then you should
' be able to write any program you want to connect to R/3
'----------------------------------------------------------------
'****************************************************************
'****************************************************************
' Declarations
'****************************************************************
'----------------------------------------------------------------
' Global LogonControl As SAPLogonCtrl.SAPLogonControl
'----------------------------------------------------------------
DIM LogonControl
'----------------------------------------------------------------
' Global conn As SAPLogonCtrl.Connection
'----------------------------------------------------------------
DIM conn
'----------------------------------------------------------------
' Global funcControl As SAPFunctionsOCX.SAPFunctions
'----------------------------------------------------------------
DIM funcControl
'----------------------------------------------------------------
' Global TableFactoryCtrl As SAPTableFactoryCtrl.SAPTableFactory
'----------------------------------------------------------------
DIM TableFactoryCtrl
'----------------------------------------------------------------
' Pointer to functions
'----------------------------------------------------------------
DIM RFC_READ_TABLE
'----------------------------------------------------------------
' Pointers to function parameters
'----------------------------------------------------------------
DIM eQUERY_TAB
DIM TOPTIONS
DIM TDATA
DIM TFIELDS
```

```
'***************************************************************
' Main Program
'***************************************************************
'---------------------------------------------------------------
   call Main
'---------------------------------------------------------------
'***************************************************************
' Subroutines
'***************************************************************
Sub Main()
'---------------------------------------------------------------
' Main is the principle Entry to the program pool
'---------------------------------------------------------------
   Set LogonControl = CreateObject("SAP.LogonControl.1")
   Set funcControl = CreateObject("SAP.Functions")
   Set TableFactoryCtrl = CreateObject("SAP.TableFactory.1")

   call R3Logon
   funcControl.Connection = conn
   call R3RFC_READ_TABLE("T000")
   conn.Logoff
   MsgBox " Logged off from R/3! "
End Sub

Sub R3Logon()
'---------------------------------------------------------------
' Logon to your R/3 system
'---------------------------------------------------------------
   Set conn = LogonControl.NewConnection
MsgBox "*** Attention. You must enter your own system data here first"
STOP
'---------------------------------------------------------------
' ** Set here your system data. They are also found in the R/3 Logon
Panel
' Only the app server and the system number is mandatory. If the other
params
' are missing you will be prompted for
'---------------------------------------------------------------
   conn.ApplicationServer = "r3dev" ' IP or DNS-Name of the R/3
application server
   conn.System = "00"               ' System ID of the instance, usually
00
   conn.Client = "100"              ' opt. Client number to logon to
   conn.Language = "EN"             ' opt. Your login language
   conn.User = ""                   ' opt. Your user id
   conn.Password = ""               ' opt. Your password

   retcd = conn.Logon(0, False)
   If retcd <> True Then
     MsgBox " Cannot log on! "
     MsgBox retcd
     Stop
   else
     MsgBox " Logon OK."
   End If
End Sub
```

```
Sub R3RFC_READ_TABLE(pQueryTab)
'-------------------------------------------------------------
' Call the R/3 RFC function RFC_READ_TABLE
'-------------------------------------------------------------
  Set RFC_READ_TABLE = funcControl.Add("RFC_READ_TABLE")
'-------------------------------------------------------------
' If you get the error message "Collection Member Not Found"
'    then you have probable misspelled the parameter name
'    e.g. you wrote QUERY_TAB instead of QUERY_TABLE
'-------------------------------------------------------------
  Set eQUERY_TAB = RFC_READ_TABLE.Exports("QUERY_TABLE")
  Set TOPTIONS   = RFC_READ_TABLE.Tables("OPTIONS")   '
  Set TDATA      = RFC_READ_TABLE.Tables("DATA")      '
  Set TFIELDS    = RFC_READ_TABLE.Tables("FIELDS")    '

  eQUERY_TAB.Value = pQueryTab
  TOPTIONS.AppendRow ' new item line
  TOPTIONS(1,"TEXT") = "MANDT EQ '000'"

  If RFC_READ_TABLE.Call = True Then
    If TDATA.RowCount > 0 Then
      MsgBox "Call to RFC_READ_TABLE successful! Data found"
      MsgBox TDATA(1, "WA")
    Else
      MsgBox "Call to RFC_READ_TABLE successful! No data found"
    End If
  Else
    MsgBox "Call to RFC_READ_TABLE failed!"
  End If
End Sub
```

Program 13:   Another Visual Basic Script Example

12.5 R/3 RFC/OLE Troubleshooting

Problems connecting via RFC can usually be solved by reinstalling the full SAPGUI and/or checking your network connection with R/3.

Reinstall the full SAPGUI

If you have problems to connect to R/3 via the RFC DLLs then you should check your network installation. It would be out of the reach of this publication to detail the causes and solutions when an RFC connection does not work.

I may say, that in most cases a full install of the SAPGUI on the computer which runs the calling program will secure a reliable connection, provided that you can login to R/3 problem-free with this very same SAPGUI installation.

Another trivial but often cause are simple network problems. So impossible it may appear, you should always go by the book and first check the network connection by pinging the R/3 system with the PING utility and checking the proper access authorities.

Check spelling

However, if you successfully passed the SAPlogon method, then the problem is mostly a misspelling of object or method names or an incompatibility of the called function.

Make certain that the function module in R/3 is marked as "RFC allowed"

If you are quite sure that you spelled everything right and correct, and still get an error executing the SAP.FUNCTIONS.CALL method then you should investigate the function module in R/3.

Check for syntax errors

Generate the function group to see if there is an syntax error
Make sure that the function is tagged as RFC allowed

13

ALE - Application Link Enabling

ALE is an R/3 technology for distribution of data between independent R/3 installations. ALE is an application which is built on top of the IDoc engine. It simply adds some structured way to give R/3 a methodical mean to find sender, receiver and triggering events for distribution data.

Make Use of ALE for Your Developments

- Transfer master data for material, customer, supplier and more to a different client or system with *BALE*

- Copy your settings for the R/3 classification and variant configurator to another system, also in *BALE*

- Copy pricing conditions with ALE from the conditions overview screen (e.g. *VV12*)

13.1 A Distribution Scenario Based On IDocs

ALE has become very famous in business circles. While it sounds mysterious and like a genial solution, it is simply a mean to automate data exchange between SAP systems. It is mainly meant to distribute data from one SAP system to the next. ALE is a mere enhancement of SAP-EDI and SAP-RFC technology.

ALE is an SAP designed concept to automatically distribute and replicate data between webbed and mutually trusting systems

Imagine your company has several sister companies in different countries. Each company uses its own local SAP installation. When one company creates master data eg. material or customer master it is much likely that these data should be known to all associates. ALE allows to immediately trigger an IDoc sent to all associates as soon as the master record is created in one system.

Another common scenario is, that a company uses different installations for company accounting and production and sales. In that case ALE allows you to copy the invoices created in SD immediately to the accounting installation.

ALE defines the logic and the triggering events who describe how and when IDocs are exchanged between the systems

ALE defines a set of database entries, which are called the ALE scenario. These tables contain the information which IDocs shall be automatically replicated to one or more connected R/3-compatible data systems.

ALE is an application put upon the IDoc and RFC mechanisms of SAP

To be clear: ALE is not a new technology. It is only a handful of customizing settings and background routines that allow timed and triggered distribution of data to and from SAP or RFC-compliant systems. ALE is thus a mere enhancement of SAP-EDI and SAP-RFC technology.

13.2 Example ALE Distribution Scenario

To better understand let us model a small example ALE scenario for distribution of master data between several offices.

Let as assume that we want to distribute three types of master data objects, the material master, the creditor master and the debtor master.

Let us assume that we have four offices. This graphic scenario shows the type of data exchanged between the offices. Any of these offices operates an own stand alone R/3 system. Data is exchanged as IDocs which are sent from the sending office and received from the receiving office.

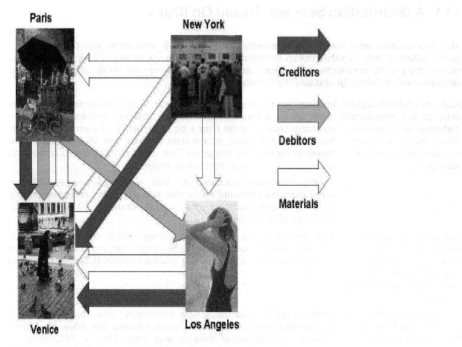

Illustration 20: **ALE distribution scenario**

Data Object		Sender		Receiver	
MATMAS	Material Master	R3NYX	New York Office	R3VEN	Venice Office
MATMAS	Material Master	R3NYX	New York Office	R3PAR	Paris Office
MATMAS	Material Master	R3NYX	New York Office	R3LAX	Los Angeles
MATMAS	Material Master	R3PAR	Paris Office	R3VEN	Venice Office
MATMAS	Material Master	R3LAX	Los Angeles	R3VEN	Venice Office
DEBMAS	Debitor Master	R3PAR	Paris Office	R3VEN	Venice Office
DEBMAS	Debitor Master	R3PAR	Paris Office	R3LAX	Los Angeles
CREMAS	Creditor Master	R3NYX	New York Office	R3VEN	Venice Office
CREMAS	Creditor Master	R3PAR	Paris Office	R3VEN	Venice Office
CREMAS	Creditor Master	R3LAX	Los Angeles	R3VEN	Venice Office

Illustration 21: **Scenario in tabular form**

13.3 How ALE Works

ALE is a simple add-on application propped upon the IDoc concept of SAP R/3. It consists of a couple of predefined ABAPs which rely on a customised distribution scenario. These scenarios simply define the IDoc types and the relationship and technical details of the partners which exchange data.

ALE defines when and how data is replicated between systems

ALE defines the logic and the triggering events which describe how and when IDocs are exchanged between the systems. If the ALE engine has determined which data to distribute, it will call an appropriate routine to create an IDoc. The actual distribution is then performed by the IDoc layer.

The predefined distribution ABAPs can be used as templates for own development

ALE is of course not restricted to the data types which are already predefined in the BALE transaction. You can write your ALE distribution handlers, which should only comply with some formal standards, e.g. not bypassing the ALE scenarios.

ALE uses IDocs to transmit data between systems

All ALE distribution uses IDocs to replicate the data to the target system. The ALE applications check with the distribution scenario and do nothing more than calling the matching IDoc function module, which is alone responsible for gathering the requested data and bringing them to the required data port. You need to thoroughly understand the IDoc concept of SAP beforehand, in order to understand ALE

The actual process is extremely simple: Every time a data object changes, which is a member of an ALE scenario, then an IDoc is triggered by one of the defined triggering mechanisms. The triggers are usually an explicit ABAP or a technical workflow event.

ABAPs can be used in batch routine

Distribution ABAPs are started manually or can be set up as a triggered or timed batch job.. Sample ABAPs for ALE distribution are those used for master data distribution in transaction BALE, like the ones behind the transaction BD10, BD12 etc.

Workflow is triggered from change document

The workflow for ALE is based on change pointers. Change pointers are entries in a special database entity, which record the creation or modification of a database object. These change pointers are very much like the SAP change documents. They are also written from within a change document, i.e. from the `function CHANGEDOCUMENT_CLOSE`. The workflow is also triggered from within this function.

Relevance for change pointers is defined in IMG

SAP writes ALE change pointers to circumvent a major disadvantage of the change documents. Change documents are only written, if a value of a table field changes and the data element of the table field is marked as relevant for change documents (see SE11). ALE change pointers use a customized table instead, which contains the names of those table fields, which shall be regarded as relevant changes.

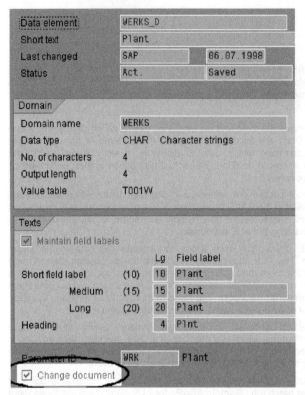

Illustration 22:   **Flag that indicates that data element writes a change document**

13.4 Useful ALE Transaction Codes

ALE is customized via three main transaction. These are *SALE*, *WEDI* and *BALE*.

***SALE* is the core transaction for ALE**

Here you find everything ALE related, which is not already covered by the other customizing transactions.

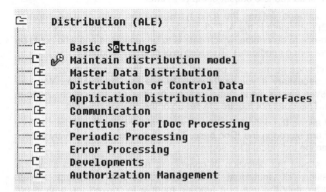

Illustration 23: **SALE - ALE Specific customizing**

***WEDI* - IDoc Administration**

Here you define all the IDoc related parts, which make up most of the work related to ALE.

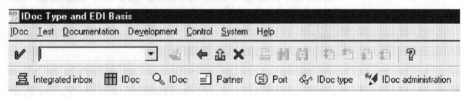

Illustration 24: **WEDI – Master menu for all EDI activities**

***BALE* – Central menu**

This is a menu, which combines most function necessary for ALE distribution, especially the triggering of manual distribution of master data or variant configuration or classification.

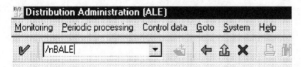

Illustration 25: **BALE – Master menu for ALE specific routine tasks**

***BDBG* - Automatically generate IDocs from a BAPI**

Good stuff for power developers. It allows to generate all IDoc definitions including segments and IDoc types from the DDIC entries for a BAPI definition.

13.5 ALE Customizing *SALE*

ALE customizing is relatively straight forward. The only mandatory task is the definition of the ALE distribution scenario.

SALE

All ALE special customizing is done from within the transaction *SALE*, which links you to a subset of the SAP IMG.

Distribution Scenarios

The scenario defines the IDoc types and the pairs of IDoc partners which participate in the ALE distribution. The distribution scenario is the reference to determine, which data is to be replicated and who could be the receiving candidates. This step is of course mandatory.

Change Pointers

The change pointers can be used to trigger the ALE distribution. This is only necessary if you really want to use that mechanism. You can however always send out IDocs every time an application changes data. This does not require the set-up of the change pointers.

Filters

SAP allows the definition of rules, which allow a filtering of data, before they are stored in the IDoc base. This allows you to selectively accept or decline individual IDoc segments.

Conversion

ALE allows the definition of conversion rules. These rules allow the transition of individual field data according mapping tables. Unfortunately the use of a function module to convert the data is not realized in the current R/3 release.

Conversion

The filter and conversion functionality is only attractive on a first glance. Form practical experience we can state, that they are not really helpful. It takes a long time to set up the rules and yet are rules usually not powerful enough to avoid modifications in an individual scenario. Conversion rules tend to remain stable, after they have once been defined. Thus it is usually easier to call am individual IDoc processing function module, which performs your desired task more flexible and easier.

13.6 Basic Settings *SALE*

Basic settings have do be adjusted before you can start working with ALE.

Logical System   Before we start we need to maintain some logical systems. These are names for the RFC destinations that are used as communication partners. An entry for each logical system is created in the table TBDLS. You can also edit the table using *SM31*.

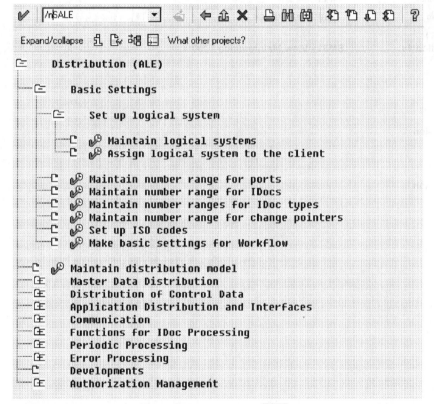

Illustration 16:	Customizing transaction *SALE*

LogSystem	Bezeichnung
TESTSENDER	ALE-Test: Logical System Sender
TESTTARGET	ALE-Test: Logical System Receiver

Illustration 17:	SM31 - View Maintenance TBDLS

http://idocs.de http://logosworld.de

ALE - Application Link Enabling

Assign logical system to a client   You will finally have to assign a logical system to the clients involved in ALE or IDoc distribution. This is done in table T000, which can be edited via SM31 or via the respective SALE tree element.

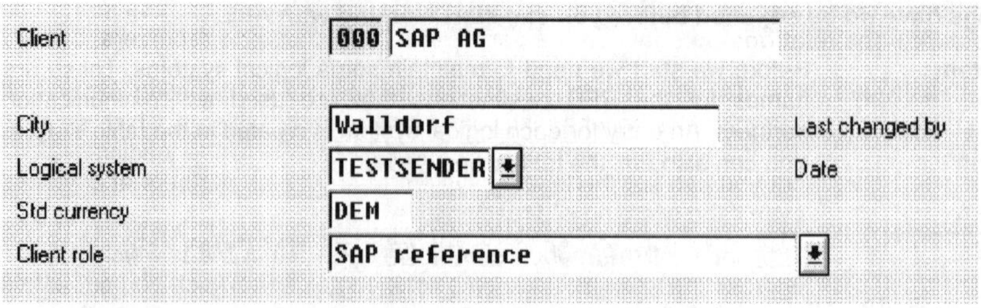

Client	000 SAP AG	
City	Walldorf	Last changed by
Logical system	TESTSENDER ±	Date
Std currency	DEM	
Client role	SAP reference ±	

Illustration 18:   **SM31 - View Maintenance T000**

13.7 Define The Distribution Model (The "Scenario") *BD64*

The distribution model (also referred to as ALE-Scenario) is a more or less graphical approach to define the relationship between the participating senders and receivers.

BD64 - Model can only be maintained by leading system

The distribution model is shared between all participating partners. It can therefore only be maintained in one of the systems, which we shall call the **leading system**. Only one system can be the leading system, but you can set the leading system to any of the partners at any time, even if the scenario is already active.

The name of the model view will be the name under which you will address the scenario. It serves as a container in which you put all the from-to relations.

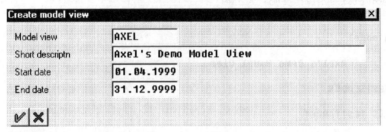

Illustration 19:   **Create an ALE model view for a scenario**

Suggestion: One scenario per administration area

You can have many scenarios for eventual different purposes. You may also want to put everything in a single scenario. As a rule of thumb it proved as successful, that you create one scenario per administrator. If you have only one ALE administrator, there is no use of having more than one scenario. If you have several departments with different requirements, that it might be helpful to create one scenario per department.

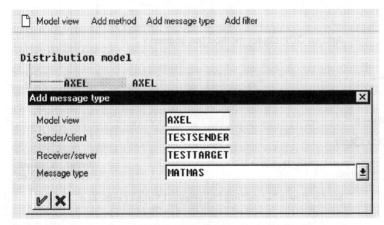

Illustration 20:   **Add a message type to the scenario**

http://idocs.de http://logosworld.de

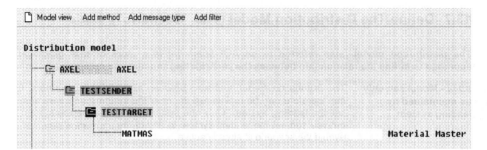

Illustration 21:   **Model View After Adding MATMAS**

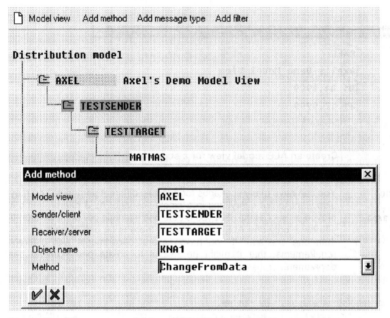

Illustration 22:   **Add an OOP object method the scenario**

Illustration 23:   **Model view after adding Customer.ChangeFromData**

13.8 Generating Partner Profiles *WE20*

A very useful utility is the automatic generation of partner profiles out of the ALE scenario. Even if you do not use ALE in your installation, it could be only helpful to define the EDI partners as ALE scenario partners and generate the partner profiles.

Now go on defining partner profiles

The model view display graphically the from-to relations between logical systems. You now have to generate the partner profiles which are used to identify the physical means of data transportation between the partners.

WE20

If you define the first profile for a partner, you have to create the profile header first. Click an the blank paper sheet.

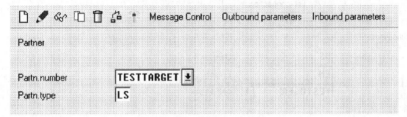

Illustration 24:   **Create a partner**

Partner class is for your own documentation

The values given for partner class are for categorization only, e.g. to identify those partners which belong to the same processing group and it is only a classification value. You can give an arbitrary name in order to group the type of partners, e.g. EDI for external ones, ALE for internal ones and IBM for connection with IBM OS/390 systems.

Partner status to activate and deactivate the profile

The partner status instead tells whether the partner definition is active. This allows you to temporarily deactivate a partner definition without having to delete the set of information.

Receiver notification sends workitem to inbox if error occurs

The **receiver notification** allows you to specify a SAP user name or a workflow role. The receiver specified in there will receive in its SAP Inbox a workitem notification every time an IDoc could not be processed successfully, .i.e. if the processing function module returned a `workflow_result` not equal to zero.

http://idocs.de http://logosworld.de

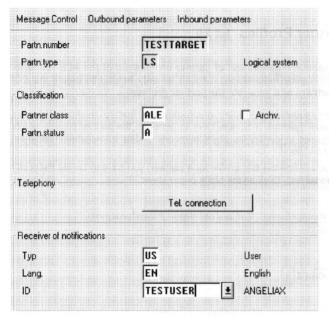

Illustration 25:   **Specify partner details**

Individual profile for each message type and partner   For every IDoc message there needs to be an individual setting defined. The values define a full technical profile for the data exchange with the partner. You may specify if data is handled immediately and which data port is used.

***SM59* to define the data ports**   The data port is the physical handler to transfer data between R/3 and an external system. Ports are defined system wide using transaction *SM59*.

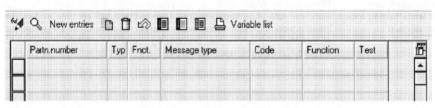

Illustration 26:   **Outbound partner profile before generation**

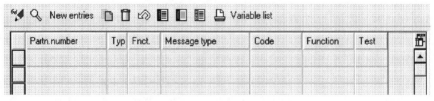

Illustration 27:   **Inbound partner profile before generation**

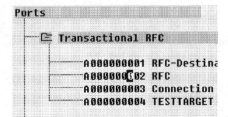

Illustration 28: **Ports defined with SM59**

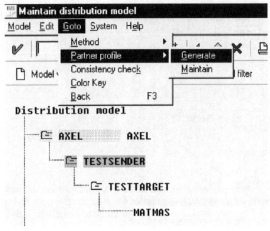

Illustration 29: **Generate Partner Profiles Form SALE menu**

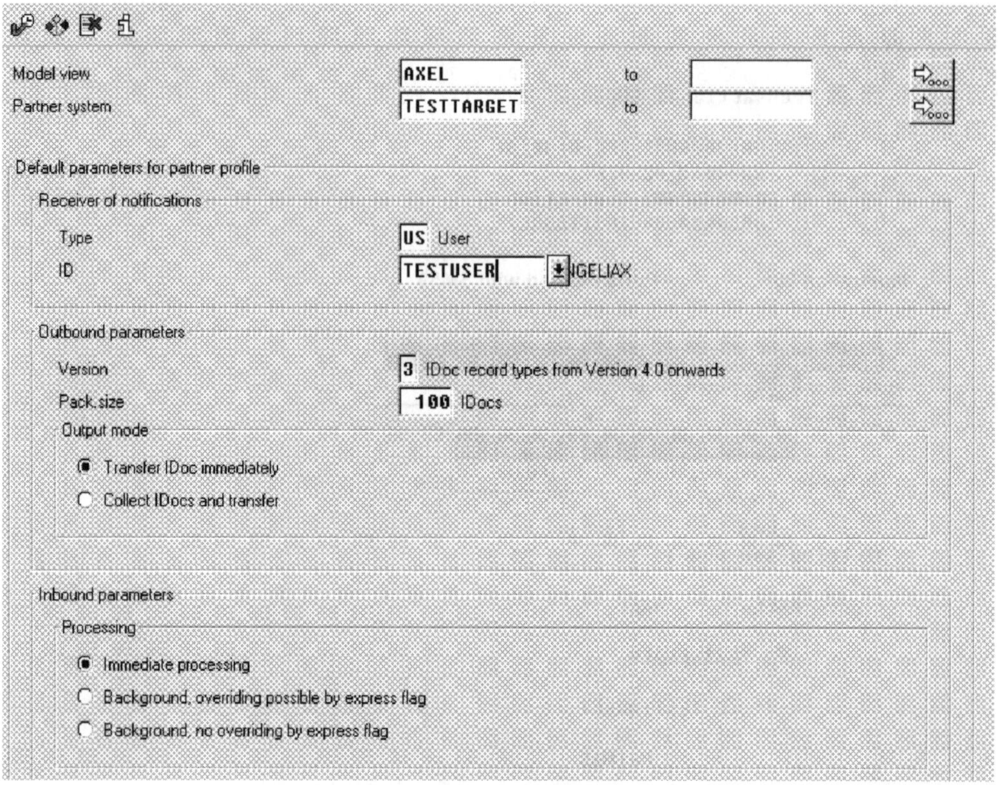

Illustration 30:   **Automatically created partner profile**

In the example have been two profiles generated. The first one is for MATMAS, which we explicitly assigned in the distribution scenario. The second one is a mandatory IDoc type with the name SYNCH which is used for RFC control information and synchronisation.

Partn.number	Typ	Fnct.	Message type	Code	Function	Test
TESTTARGET	LS		MATMAS			☐
TESTTARGET	LS		SYNCH			☐

Illustration 31:   **Outbound partner profile after generation**

Here is a detail view of the parameters generated. The receiver port is the RFC destination, that had been created for TESTTARGET with SM59.
Data goes to table EDP13.

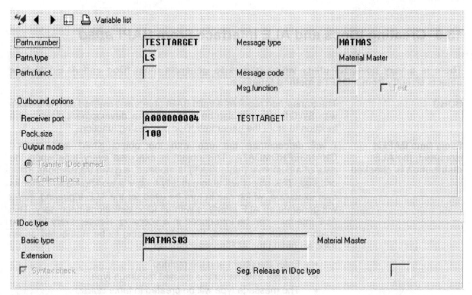

Illustration 32: **Assigning the port to partner link**

13.9 Creating IDocs and ALE Interface From BAPI *SDBG*

There is a very powerful utility which allows to generate most IDoc and ALE interface objects directly from a BAPI's method interface.

BDBG

The transaction requires a valid BAPI object and method as it is defined with SWO1. You will also have to specify a development class and a function to store the generated IDoc processing function.

Every time BAPI is executed, the ALE distribution is checked

I will demonstrate the use with the object KNA1 and method CHANGEFROMDATA. This object is executed every time when the data of a customer (table KNA1) is modified, e.g. via transactions *XD01* or *XD02*. This object will automatically trigger a workflow event after its own execution, which can in turn be used for the ALE triggering. *BDBG* will generate an ALE interface with all IDoc definitions necessary. This ALE interface can be introduced in a scenario. Hence, every time the customer data is modified, the data is going to be distributed as an IDoc according the ALE scenario set-up.

- Enter the object and the method.
- Specify a name for the created message type.
- The message type will be created in table EDMSG .
- Define the names of the processing function modules and the associated IDoc types.

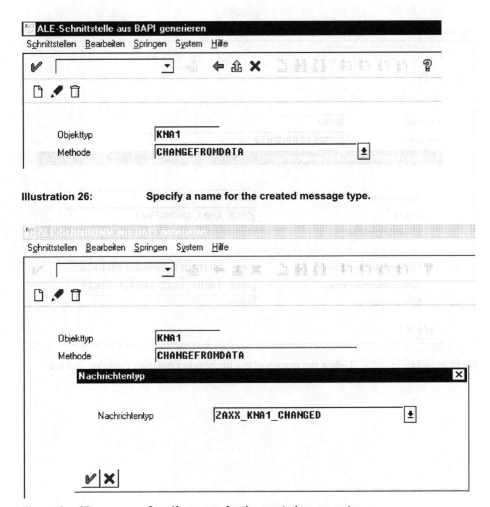

Illustration 26:   **Specify a name for the created message type.**

Illustration 27:   **Specify a name for the created message type**

Now you can specify the required IDoc types and the names of the
function module and function group for the processing routines. Note,
that the development class (Entwicklungsklasse) and the function
group (Funktionsgruppe) need to be in your customer name space, i.e.
should begin with Y or Z. The values proposed on this screen are
usually inappropriate.

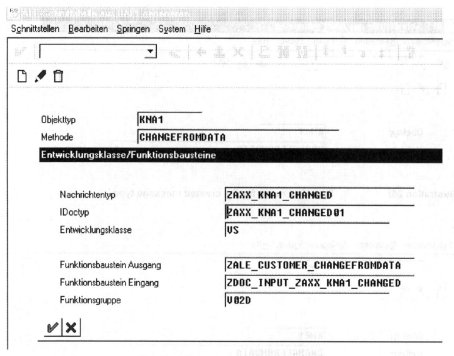

Illustration 28: **Define the names of the processing function modules and the associated IDoc types**

Result report   Click on generated objects to see what was generated in detail

```
ALE-Schnittstelle aus BAPI generieren
----------------------------------------------------------

Nachrichtentyp
   ZAXX_KNA1_CHANGED
      ZAXX_KNA1_CHANGED wurde erfolgreich generiert

IDoctyp
   ZAXX_KNA1_CHANGED01
      Prüfung des Basistyps ZAXX_KNA1_CHANGED01
      Der Basistyp ZAXX_KNA1_CHANGED01 ist mit der logis
      Es existiert kein Vorgänger
      Basistyp ZAXX_KNA1_CHANGED01 ist nicht freigegeben

Segment
   Z1ZAXX_KNA1_CHANGED
      Z1ZAXX_KNA1_CHANGED wurde erfolgreich generiert
   Z1BPKNA101
      Z1BPKNA101 wurde erfolgreich generiert

Funktionsbaustein für ALE-Ausgang
   ZALE_CUSTOMER_CHANGEFROMDATA
      ZALE_CUSTOMER_CHANGEFROMDATA wurde erfolgreich gen

Funktionsbaustein für ALE-Eingang
   ZDOC_INPUT_ZAXX_KNA1_CHANGED
      ZDOC_INPUT_ZAXX_KNA1_CHANGED wurde erfolgreich gen
```

Illustration 33:   **Generation protocol**

A detailed report is shown. The report is clickable so that you can directly view the generated objects

The transaction has generated an IDoc type with a header section containing the interface values of the object and a data section with the remaining fields of the object data structure. All information has been taken from the BAPI's interface definition.

```
FUNCTION bapi_customer_changefromdata.
*"----------------------------------------------------------------
*"*"Lokale Schnittstelle:
*"  IMPORTING
*"           VALUE(PI_ADDRESS)      LIKE BAPIKNA101 STRUCTURE BAPIKNA101
*"           VALUE(PI_SALESORG)     LIKE BAPIKNA102-SALESORG
*"           VALUE(PI_DISTR_CHAN)   LIKE BAPIKNA102-DISTR_CHAN OPTIONAL
*"           VALUE(PI_DIVISION)     LIKE BAPIKNA102-DIVISION OPTIONAL
*"           VALUE(CUSTOMERNO)      LIKE BAPIKNA103-CUSTOMER
*"  EXPORTING
*"           VALUE(PE_ADDRESS)      LIKE BAPIKNA101 STRUCTURE BAPIKNA101
*"           VALUE(RETURN)          LIKE BAPIRETURN STRUCTURE BAPIRETURN
*"----------------------------------------------------------------
```

Illustration 34: **Function interface of the BAPI used to generate the IDoc interface**

Generated segment structure from BAPI function interface parameter

For each of the parameters in the BAPIs interface, the generator created a segment for the IDoc type. Some segments are used for IDoc inbound only, others for IDoc outbound instead. Parameter fields that are not structured will be combined in a single segment which is placed as first segment of the IDoc type and contains all these fields. This collection segment receives the name of the IDoc type. In our example this is the generated segment Z1ZAXX_KNA1_CHANGED.

The segment below has been created as a header level segment and combines all function module parameters, which do not have a structure, i.e. which are single fields. E.g. if the BAPI has parameters a parameter like "i_material LIKE mara-matnr" then it will be placed in the control segment. However if it is declared like "i_material STRUCTURE mara" then it will create an own IDoc segment.

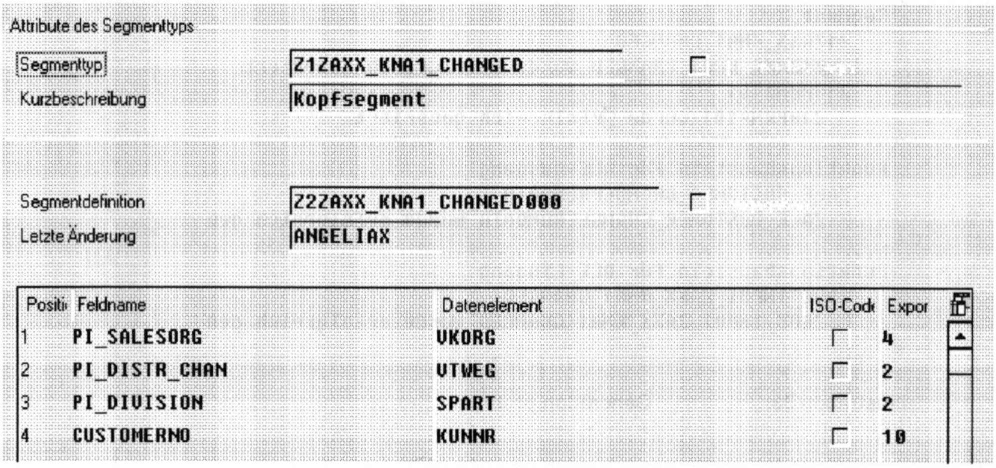

Attribute des Segmenttyps

Segmenttyp	Z1ZAXX_KNA1_CHANGED
Kurzbeschreibung	Kopfsegment

Segmentdefinition	Z2ZAXX_KNA1_CHANGED000
Letzte Änderung	ANGELIAX

Positi	Feldname	Datenelement	ISO-Cod	Expor
1	PI_SALESORG	VKORG		4
2	PI_DISTR_CHAN	VTWEG		2
3	PI_DIVISION	SPART		2
4	CUSTOMERNO	KUNNR		10

Illustration 35: **Segment Z1ZAXX_KNA1_CHANGED**

13.10 Defining Filter Rules

ALE allows to define simple filter and transformation rules. These are table entries, which are processed every time the IDoc is handed over to the port. Depending on the assigned path this happens either on inbound or outbound.

SALE

Rules are defined with the *SALE* transaction.

Conversion rules are less powerful than programmed routines

The filter and conversion functionality is only attractive on a first glance. From practical experience we can state, that they are not really helpful. It takes a long time to set up the rules and yet are rules usually not powerful enough to avoid modifications in an individual scenario. Conversion rules tend to remain stable, after they have once been defined. Thus it is usually easier to call an individual IDoc processing function module, which performs your desired task more flexible and easier.

It is cheaper to find an error in ABAP than in ALE customizing

It is generally cheaper, faster and easier to maintain, if you avoid setting up rules and rewrite own processing routines. The reason is mainly, that in a productive environment you usually have at least one administrator available, who is faintly familiar with ABAP and then can search and debug the ABAP code if something goes awry, while even experienced ALE experts usually forget pretty soon, how the rules are set up.

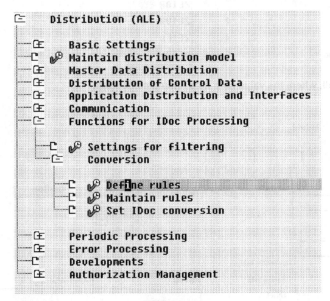

Illustration 36: *SALE*

Conversion rule	Description	IDOC segment name	
MATNR		E1MARAM	

Illustration 37: Assigning the conversion rule to an IDoc segment

Q 🖨 🗑 📋 Create proposal for rule

	Rec. field	Descript.	Type	Length	Sender fld
	MSGFN	Function	C	3	MSGFN
	MATNR	Material	C	18	MATNR
	ERSDA	Created on	C	8	ERSDA
	ERNAM	Created by	C	12	ERNAM
	LAEDA	Last change	C	8	LAEDA
	AENAM	Changed by	C	12	AENAM
	PSTAT	Maint. status	C	15	PSTAT
	LVORM	DF client level	C	1	LVORM
	MTART	Material type	C	4	MTART
	MBRSH	Industry sector	C	1	MBRSH
	MATKL	Material group	C	9	MATKL
	BISMT	Old matl number	C	18	BISMT
	MEINS	Base unit	C	3	MEINS
	BSTME	Order unit	C	3	BSTME
	ZEINR	Document	C	22	ZEINR
	ZEIAR	Document type	C	3	ZEIAR
	ZEIVR	Doc. version	C	2	ZEIVR
	ZEIFO	Page format	C	4	ZEIFO
	AESZN	Doc. change no.	C	6	AESZN

Illustration 38:   **Tell, where the value for a field should come fromt**

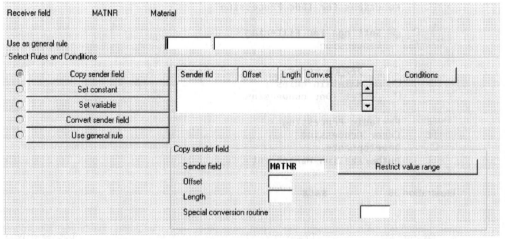

Illustration 39:   **Define a rule**

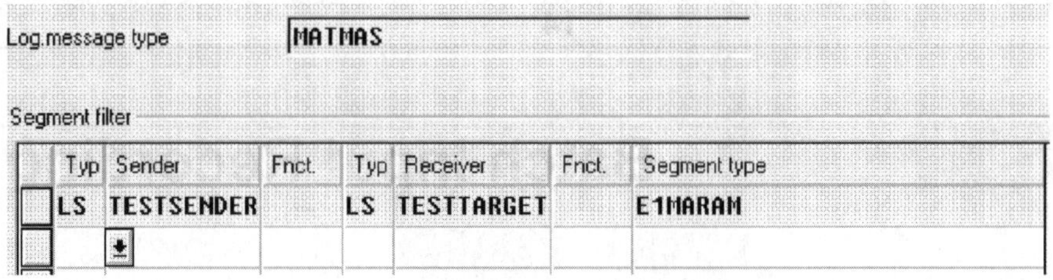

	Typ	Sender	Fnct.	Typ	Receiver	Fnct.	Segment type
Log.message type		MATMAS					

Segment filter

	Typ	Sender	Fnct.	Typ	Receiver	Fnct.	Segment type
	LS	TESTSENDER		LS	TESTTARGET		E1MARAM
		⬇					

Illustration 40:   **Assigning the filter to a partner link**

Batch Input Recording

The batch input (BTCI) recorder (*SHDB*) is a precious tool to develop inbound IDocs. It records any transaction like a macro recorder. From the recording an ABAP fragment can be created. This lets you easily create data input programs, without coding new transactions.

14.1 Recording a Transaction With *SHDB*

The BTCI recorder lets you record the screen sequences and values entered during a transaction. It is one of the most precious tools in R/3 since release 3.1. It allows a fruitful cooperation between programmer and application consultant.

The section below will show you an example of, how the transaction *SHDB* works. With the recording you can easily create an ABAP, which is able to create BTCI files.

Record a session with transaction SHDB

You will be asked for a session name and the name of the transaction to record. Then you can enter the data into the transaction as usual.

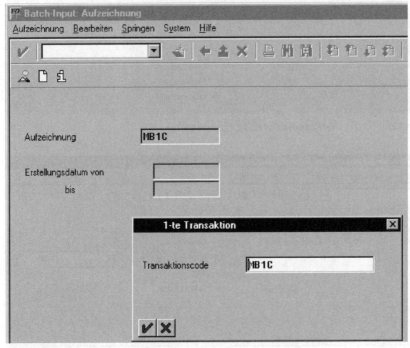

Illustration 41:   **Starting a new recording with SHDB**

Now the transaction is played and all entries recorded

The following screens will show the usual transaction screens. All entries that you make are recorded together with the screen name and eventual cursor positions.

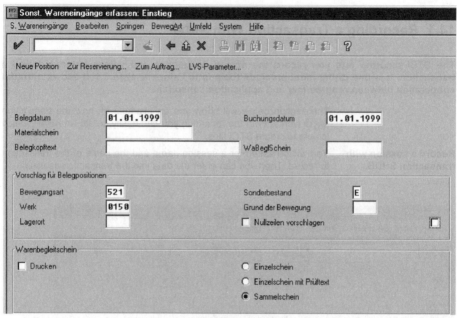

Illustration 42:   First screen of `MB1C` (goods entry)

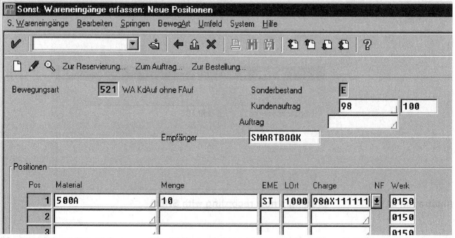

Illustration 43:   Recorded list screen for goods entry

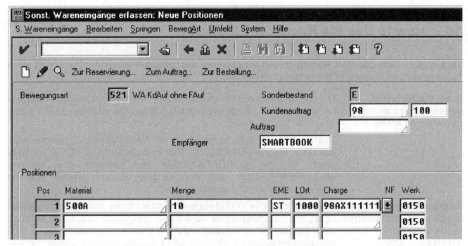

Illustration 44:   **Recorded Detail Screen for goods entry**

From the recorded session, you can generate an ABAP

After you finished the recording you have the possibility to generate ABAP coding from it. This will be a sequence of statements which can generate a batch input session, which is an exact replay of the recorded one.

The generated program contains an include BDCRECXX which contains all the FORM routines referenced.

Put the coding into a function module

To make the recorded code usable for other program, you should make a function module out of it. Start9ing with release 4.5A the recorded provides a feature to automatically generate such a function module. For earlier release we give the coding of a program which fulfils this task further down.

14.2 How to Use the Recorder Efficiently

This routine replaces BDCRECXX to allow executing the program generated by SHDB via a call transaction instead of generating a BTCI file.

From the recorded session, you can generate an ABAP

The SHDB transaction creates an ABAP from the recording. When you run this ABAP, it will generate a BTCI group file, with exactly the same data as in the recording.

The recorder is able to generate an ABAP. Releases before 4.5A include a routine BDCRECXX. This include contains FORM routines which fill the BDCDATA table and execute the routines BDC_OPEN_GROUP and BDC_CLOSE_GROUP. These are the routines which create batch input files.

Replace the include with modified FORM routines to allow CALL TRANSACTION

If we modified this FORM routines a little bit, we can make the ABAP replay the recording online via a `CALL TRANSACTION`, which is much more suitable for our development and testing purposes. If you replace the standard include `BDCRECXX` with the shown one `ZZBDCRECXX`, you can replay the recording online.

Starting with release 4.5A you can create a function module from the recording. This function modules replace the recorded constants with parameters and give you the option to choose between a batch input file or a direct call transaction.

Scrolling areas with table controls require to modify the recording and to add a loop.

A remark on screen processing, if there are table controls (scroll areas). If you enter many lines or try to extend a list, where you do cannot tell before, how many lines the list contains, you will not know, where to place the cursor. Therefore most transactions provide a menu option, that positions the list in a calculable manner. If you choose a new item, most transaction will either pop up a detail screen or will position the list, so that the next free line is always line 2. If this feature is not provided in a transaction, it is regarded as a malfunction by SAP and can be reported to SAPNET/OSS.

14.3 Include ZZBDCRECXX to Replace BDCRECXX

This routine replaces BDCRECXX to allow executing the program generated by *SHDB* via a call transaction instead of generating a BTCI file.

```
*------------------------------------------------------------------------*
*     INCLUDE ZZBDCRECXX                                                  *
*------------------------------------------------------------------------*
FORM OPEN_GROUP.
  REFRESH BDCDATA.
ENDFORM.
*------------------------------------------------------------------------*
FORM CLOSE_GROUP.
ENDFORM.
*------------------------------------------------------------------------*
FORM BDC_TRANSACTION USING TCODE.
  CALL TRANSACTION TCODE USING BDCDATA MODE 'A' MESSAGES INTO BDCMESS.
ENDFORM.
*------------------------------------------------------------------------*
FORM BDC_TRANSACTION_MODE  USING TCODE AMODE.
  CALL TRANSACTION TCODE USING BDCDATA UPDATE 'S'
       MODE AMODE MESSAGES INTO BDCMESS.
ENDFORM.
*------------------------------------------------------------------------*
FORM BDC_DYNPRO USING PROGRAM DYNPRO.
  CLEAR BDCDATA.
  BDCDATA-PROGRAM  = PROGRAM.
  BDCDATA-DYNPRO   = DYNPRO.
  BDCDATA-DYNBEGIN = 'X'.
  APPEND BDCDATA.
ENDFORM.
*------------------------------------------------------------------------*
FORM BDC_FIELD USING  FNAM FVAL.
  FIELD-SYMBOLS: <FLD>.
  ASSIGN (FNAM) TO <FLD>.
  CLEAR BDCDATA.
  DESCRIBE FIELD FVAL TYPE SY-FTYPE.
  CASE SY-FTYPE.
    WHEN 'C'.
       WRITE FVAL TO BDCDATA-FVAL.
    WHEN OTHERS.
       CONDENSE FVAL.
       WRITE FVAL TO BDCDATA-FVAL LEFT-JUSTIFIED.
  ENDCASE.
  BDCDATA-FNAM = FNAM.
  APPEND BDCDATA.
ENDFORM.                              " BDC_FIELD
```

```
*-----------------------------------------------------------------------*
FORM GET_MESSAGES TABLES    P_MESSTAB STRUCTURE BDCMSGCOLL.
  P_MESSTAB[] = BDCMESS[].
  LOOP AT P_MESSTAB.
    AT LAST.
      READ TABLE P_MESSTAB INDEX SY-TABIX.
      MOVE-CORRESPONDING P_MESSTAB TO SYST.
    ENDAT.
  ENDLOOP.
ENDFORM.                                  " GET_MESSAGES
*-----------------------------------------------------------------------*
FORM GET_RESULTS TABLES  MESSTAB STRUCTURE BDCMSGCOLL
                         RETURN_VARIABLES STRUCTURE BDWFRETVAR
               CHANGING WORKFLOW_RESULT LIKE BDWF_PARAM-RESULT.
  PERFORM GET_MESSAGES TABLES MESSTAB.
  DESCRIBE TABLE MESSTAB LINES SY-TFILL.
  REFRESH: RETURN_VARIABLES.
  CLEAR: WORKFLOW_RESULT, RETURN_VARIABLES.
  WORKFLOW_RESULT = 99999.
  IF SY-TFILL GT 0.
    READ TABLE MESSTAB INDEX SY-TFILL.
    IF MESSTAB-MSGTYP CA 'S'.
      WORKFLOW_RESULT = 0.
      RETURN_VARIABLES-DOC_NUMBER = MESSTAB-MSGV1.
      APPEND RETURN_VARIABLES.
    ENDIF.
  ENDIF.
ENDFORM.                    " GET_RESULTS
```

Program 14:   Program ZZBDCRECXX (find at http://www.idocs.de)

14.4   Generate a Function from Recording

The section shows the coding of routine ZZBDCRECXX_FB_GEN . This goodie can replaces BDCRECXX in a recorded ABAP. Upon executing, it will generate a function module from the recording with all variables as parameters.

The ABAP generated by SHDB is a very useful tool for developers. However, it does not replace the recorded constants by variables.

The following routine generates a function module from the recording. All you have to do is, to put the coding below in an include.

ZZBDCRECXX_FBGEN   Give it the name ZZBDCRECXX_FBGEN.

Replace BDCRECXX   Then replace the include BDCRECXX in the recording with ZZBDCRECXX_FBGEN.

Execute the ABAP once   When you execute the ABAP, a function module in an existing function group will be created. The created function will contain the recording with all the constants replaced by variables, which show in the function module interface.

The following useful routine is written for releases up to 4.0B. In release 4.5B a similar functionality is provided. You can generate a function module from the recording transaction directly.

Before you generate the function, a function group must exist. This you have to do manually. The function group must also contain the include ZZBDCRECXX shown before, to have the declarations of the referenced FORM routines.

```
*-----------------------------------------------------------------------*
PARAMETERS: FUNCNAME LIKE RS38L-NAME DEFAULT 'Z_TESTING_BTCI_$1'.
PARAMETERS: FUGR     LIKE RS38L-AREA DEFAULT 'Z_BTCI_TESTING'.
*-----------------------------------------------------------------------*
DATA: TABAP LIKE ABAPTEXT OCCURS 0 WITH HEADER LINE.
DATA: BEGIN OF XCONST OCCURS 0,
  NAM LIKE DD03L-FIELDNAME, FREF LIKE DD03L-FIELDNAME,
  FVAL LIKE BDCDATA-FVAL,   FIDX(6),
       END OF XCONST.
DATA: STRL1 LIKE SY-FDPOS.
DATA: STRL2 LIKE STRL1.
DATA: IMPORT_PARAMETER    LIKE RSIMP    OCCURS 0 WITH HEADER LINE.
DATA: EXPORT_PARAMETER    LIKE RSEXP    OCCURS 0 WITH HEADER LINE.
DATA: TABLES_PARAMETER    LIKE RSTBL    OCCURS 0 WITH HEADER LINE.
DATA: CHANGING_PARAMETER  LIKE RSCHA    OCCURS 0 WITH HEADER LINE.
DATA: EXCEPTION_LIST      LIKE RSEXC    OCCURS 0 WITH HEADER LINE.
DATA: PARAMETER_DOCU      LIKE RSFDO    OCCURS 0 WITH HEADER LINE.
DATA: SHORT_TEXT LIKE TFTIT-STEXT
               VALUE 'Generated BTCI for transaction ##'.
DATA: XTCODE     LIKE SY-TCODE.
DATA: STR255(255).
TABLES: TLIBG, TFDIR.
*-----------------------------------------------------------------------*
FORM OPEN_GROUP.
  FORMAT COLOR COL_TOTAL.
  WRITE: / 'Trying to generate function ', FUNCNAME.
  FORMAT RESET.
  ULINE.
  SELECT SINGLE * FROM TLIBG WHERE AREA EQ FUGR.
  IF SY-SUBRC NE 0.
```

```
    MESSAGE I000(38) WITH 'Function Pool' FUGR 'does not exit'.
    EXIT.
  ENDIF.
  MOVE 'PERFORM OPEN_GROUP.' TO TABAP.
  APPEND TABAP.
*------------------------------------------------------------------------*
  XCONST-FNAM = 'INPUT_METHOD'.
  XCONST-FREF = 'BDWFAP_PAR-INPUTMETHD'.
  XCONST-FVAL = 'A'.
  APPEND XCONST.
ENDFORM.
*------------------------------------------------------------------------*
FORM CLOSE_GROUP.
  LOOP AT XCONST.
    IMPORT_PARAMETER-PARAMETER  = XCONST-FNAM.
    IMPORT_PARAMETER-DBFIELD    = XCONST-FREF.
    CONCATENATE '''' XCONST-FVAL '''' INTO
        IMPORT_PARAMETER-DEFAULT.
    IMPORT_PARAMETER-OPTIONAL    = 'X'.

    CASE XCONST-FIDX.
      WHEN 'E'.
        MOVE-CORRESPONDING IMPORT_PARAMETER TO EXPORT_PARAMETER.
        APPEND EXPORT_PARAMETER.
      WHEN '*'.
      WHEN OTHERS.
        APPEND IMPORT_PARAMETER.
    ENDCASE.

* --make table parameters for obvious loop fields (fields with index)
    IF XCONST-FIDX CA ')*'.
      MOVE-CORRESPONDING IMPORT_PARAMETER TO TABLES_PARAMETER.
      TABLES_PARAMETER-DBSTRUCT = IMPORT_PARAMETER-DBFIELD.
      IF XCONST-FIDX NE '*'.
        TABLES_PARAMETER-PARAMETER(1) = 'T'.
      ENDIF.
      IF XCONST-FIDX CA '*'.
        APPEND TABLES_PARAMETER.
      ENDIF.
      FORMAT COLOR COL_POSITIVE.
    ENDIF.
    WRITE: / XCONST-FNAM COLOR COL_TOTAL, (60) XCONST-FVAL.

  ENDLOOP.
* SORT import_parameter BY parameter.
* DELETE ADJACENT DUPLICATES FROM import_parameter COMPARING parameter.
* SORT tables_parameter BY parameter.
* DELETE ADJACENT DUPLICATES FROM tables_parameter COMPARING parameter.
*------------------------------------------------------------------------*
  LOOP AT TABAP.
    WRITE: / TABAP COLOR COL_KEY.
  ENDLOOP.
*------------------------------------------------------------------------*
  REPLACE '##' WITH XTCODE INTO SHORT_TEXT.
  WRITE: / FUNCNAME COLOR COL_NEGATIVE.
  WRITE: / SHORT_TEXT.
  SELECT SINGLE * FROM TFDIR WHERE FUNCNAME EQ FUNCNAME.
  IF SY-SUBRC EQ 0.
    MESSAGE I000(38) WITH 'Function' FUNCNAME 'already exists'.
    PERFORM SUCCESS_MESSAGE
```

```
                USING 'Function' FUNCNAME 'already exists' SPACE ' '.
      EXIT.
    ENDIF.
    CALL FUNCTION 'RPY_FUNCTIONMODULE_INSERT'
          EXPORTING
                FUNCNAME                 = FUNCNAME
                FUNCTION_POOL            = FUGR
                SHORT_TEXT               = SHORT_TEXT
          TABLES
                IMPORT_PARAMETER         = IMPORT_PARAMETER
                EXPORT_PARAMETER         = EXPORT_PARAMETER
                TABLES_PARAMETER         = TABLES_PARAMETER
                CHANGING_PARAMETER       = CHANGING_PARAMETER
                EXCEPTION_LIST           = EXCEPTION_LIST
                PARAMETER_DOCU           = PARAMETER_DOCU
                SOURCE                   = TABAP
          EXCEPTIONS
                OTHERS                   = 7.
    IF SY-SUBRC NE 0.
        MESSAGE I000(38) WITH 'Error creating' 'Function ' FUNCNAME.
    ENDIF.
ENDFORM.
*-------------------------------------------------------------------*
FORM BDC_TRANSACTION USING TCODE.
  APPEND '*'   TO TABAP.
  MOVE 'PERFORM BDC_TRANSACTION_MODE USING I_TCODE INPUT_METHOD.'
                   TO TABAP.
  APPEND TABAP.
*-------------------------------------------------------------------*
  XTCODE = TCODE.
  STR255 = FUNCNAME.
  REPLACE '$1' WITH XTCODE INTO STR255.
  CONDENSE STR255 NO-GAPS.
  FUNCNAME = STR255.
*-------------------------------------------------------------------*
  XCONST-FNAM = 'I_TCODE'.
  XCONST-FREF = 'SYST-TCODE'.
  XCONST-FVAL = TCODE.
  XCONST-FIDX = SPACE.
  INSERT XCONST INDEX 1.
*-------------------------------------------------------------------*
  MOVE 'PERFORM GET_RESULTS TABLES TMESSTAB' TO TABAP.
  APPEND TABAP.
  MOVE '                             RETURN_VARIABLES' TO TABAP.
  APPEND TABAP.
  MOVE '                   USING ''1''              ' TO TABAP.
  APPEND TABAP.
  MOVE '                  CHANGING WORKFLOW_RESULT .' TO TABAP.
  APPEND TABAP.
  MOVE ' READ TABLE RETURN_VARIABLES INDEX 1.' TO TABAP.
  APPEND TABAP.
  MOVE ' DOC_NUMBER = RETURN_VARIABLES-DOC_NUMBER.' TO TABAP.
  APPEND TABAP.
*-------------------------------------------------------------------*
  XCONST-FNAM = 'TMESSTAB'.
  XCONST-FREF = 'BDCMSGCOLL'.
  XCONST-FVAL = SPACE.
  XCONST-FIDX = '*'.
  INSERT XCONST INDEX 1.
*-------------------------------------------------------------------*
```

```
  XCONST-FNAM = 'RETURN_VARIABLES'.
  XCONST-FREF = 'BDWFRETVAR'.
  XCONST-FVAL = SPACE.
  XCONST-FIDX = '*'.
  INSERT XCONST INDEX 1.
*------------------------------------------------------------------------*
  XCONST-FNAM = 'WORKFLOW_RESULT'.
  XCONST-FREF = 'BDWF_PARAM-RESULT'.
  XCONST-FVAL = SPACE.
  XCONST-FIDX = 'E'.
  INSERT XCONST INDEX 1.
*------------------------------------------------------------------------*
  XCONST-FNAM = 'APPLICATION_VARIABLE'.
  XCONST-FREF = 'BDWF_PARAM-APPL_VAR'.
  XCONST-FIDX = 'E'.
  INSERT XCONST INDEX 1.
*------------------------------------------------------------------------*
  XCONST-FNAM = 'DOC_NUMBER'.
  XCONST-FREF = SPACE.
  XCONST-FIDX = 'E'.
  INSERT XCONST INDEX 1.
ENDFORM.
*------------------------------------------------------------------------*
FORM BDC_DYNPRO USING PROGRAM DYNPRO.
  TABAP = '*'.
  APPEND TABAP.
  CONCATENATE
      'PERFORM BDC_DYNPRO USING ''' PROGRAM '''' ' ''' DYNPRO '''.'
                              INTO TABAP.
  APPEND TABAP.
ENDFORM.
*------------------------------------------------------------------------*
FORM BDC_FIELD USING  FNAM FVAL.
  DATA: XFVAL LIKE BDCDATA-FVAL.
  CLEAR XCONST.
  CASE FNAM.
    WHEN 'BDC_OKCODE' OR 'BDC_CURSOR' OR 'BDC_SUBSCR'.
      CONCATENATE '''' FVAL '''' INTO XFVAL.
      PERFORM ADD_BDCFIELD USING FNAM XFVAL.
    WHEN OTHERS.
      SPLIT FNAM AT '(' INTO XCONST-FREF XCONST-FIDX.
      CONCATENATE 'I_' FNAM INTO XCONST-FNAM.
      TRANSLATE XCONST-FNAM USING '-_(_) '." No dashes allowed
      MOVE FVAL TO XCONST-FVAL.
      TRANSLATE XCONST-FVAL TO UPPER CASE.
      APPEND XCONST.
      PERFORM ADD_BDCFIELD USING FNAM XCONST-FNAM.
  ENDCASE.
ENDFORM.                                  " BDC_FIELD
*------------------------------------------------------------------------*
FORM ADD_BDCFIELD USING FNAM XFNAM.
  CONCATENATE
      'PERFORM BDC_FIELD USING ''' FNAM ''' ' INTO TABAP.
  STRL1 = STRLEN( TABAP ) + STRLEN( XFNAM ).
  IF STRL1 GT 76.
    APPEND TABAP.
    CLEAR TABAP.
  ENDIF.
  CONCATENATE TABAP XFNAM '.' INTO TABAP SEPARATED BY SPACE.
  APPEND TABAP.
```

```
ENDFORM.                                 " add_bdcfield usinf fnam fval.
*------------------------------------------------------------------------*
FORM SUCCESS_MESSAGE USING V1 V2 V3 V4 OK.
  CONCATENATE V1 V2 V3 V4 INTO SY-LISEL SEPARATED BY SPACE.
  REPLACE '##' WITH FUNCNAME INTO SY-LISEL.
  MODIFY LINE 1.
  IF OK EQ SPACE.
    MODIFY LINE 1 LINE FORMAT COLOR COL_NEGATIVE.
  ELSE.
    MODIFY LINE 1 LINE FORMAT COLOR COL_POSITIVE.
  ENDIF.
ENDFORM. "ccess_message USING v1 v2 v3 v4 ok.
```

Program 15:   Program ZZBDCRECXX_FBGEN found on http://www.idocs.de

Test the function module and add eventual loops for detail processing.   The created function module should work without modification for testing at least. However, you probably will need to modify it, e.g. by adding a loop for processing multiple entries in a table control (scroll area).

15

EDI and International Standards

With the growing importance of EDI the fight for international standards heats up. While there are many business sectors like the automotive industry and book distribution who use EDI for a long time and want to continue their investment, there are others who insist in a new modern standard for everybody.

The battle is still to reach its climax, but I shall estimate that the foray of the W3C for XML will succeed and make XML the EDI standard of the future

15.1 EDI and International Standards

Electronic Data Interchange (EDI) as a tool for paperless inter-company communication and basic instrument for e-commerce is heavily regulated by several international standards.

Unfortunately it is true for many areas in the industry, that an international standard does not mean, that everybody uses the same conventions.

Manifold standards result in a Babylon

Too many organizations play their own game and define standards more or less compatible with those set by competing organizations.

National organisations versus ANSI/ISO

The main contenders are the national standards organizations and private companies versus the big international organizations ISO and ANSI.

Private companies want well established standards

The private companies being backed up by their country organizations usually fight for maintaining conventions, which have been often established for many years with satisfaction.

All inclusive standards by the big ones ANSI and ISO

The big *American National Standards Organisation* ANSI and the international partner *International Standards Organization* ISO would usually fight for a solid open standard to cover the requirements of everybody.

Pragmatism beats completeness

This generally leads to a more or less foul trade-off between pragmatism and completeness . Tragically the big organizations put themselves in question. Their publications are not free of charge. The standards are publications which cost a lot of money. So the mostly remain unread.

Standards need to be accessible and published free of charge

Nowadays computing standards have mostly been published and established by private organizations who made their knowledge accessible free of charge to everybody. Examples are manifold like PostScript by Adobe, HTML and JavaScript by Netscape, Java by SUN, SCSI by APPLE, ZIP by PK Systems or MP3 by – who cares, XML by W3C and EDIFACT by the United Nations Organization UNESCO.

15.2 Characteristics of the Standards

The well-known standards EDIFACT, X.12 and XML have similar characteristics and are designed like a document description language. Other standards and R/3 IDocs are based on segmented files.

ANSI X.12

ANSI X.12 is the US standard for EDI and e-commerce. Why, still it is. There are chances that X.12 will be soon replaced by the more flexible XML, especially with the upcoming boost of e-commerce. ANSI X.12 is a document description language.

An ANSI X.12 message is made up of segments with fields. The segments have a segment identifier and the fields are separated by a special separator character, e.g. an asterisk.

```
BEG*00*NE*123456789**991125**AC~
```

EDIFACT/UN

EDIFACT has originally been a European standard. It became popular when being chosen by the UNO for their EDI transactions. EDIFACT is a document description language. EDIFACT is very similar to ANSI X.12 and differs merely in syntactical details and the meaning of tags.

XML

XML and the internet page description language HTML are both subsets derived from the super standard SGML...

The patent and trademark holder of XML (W3C, http://w3c.org) describes the advantages of XML very precisely as follows.

1. XML is a method for putting structured data in a text file
2. XML looks a bit like HTML but isn't HTML
3. XML is text, but isn't meant to be read
4. XML is verbose, but that is not a problem
5. XML is license-free and platform-independent

And XML is fully integrated in the world wide web. It can be said briefly: XML sends the form just as the customer entered the data.

15.3 XML

This is an excerpt of an XML EDI message. The difference to all other EDI standards is, that the message information is tagged in a way, that it can be displayed in human readable form by a browser.

XML differs from the other standards. It is a document markup language like its sister and subset HTML.

XML defines additional tags to HTML, which are specially designed to mark up formatted data information.

The advantage is, that the XML message has the same information as an EDIFACT or X.12 message. In addition it can be displayed in an XML capable web browser

```
<!DOCTYPE Sales-Order PUBLIC>
<Purchase Order Customer="123456789" Send-
to="http://www.idocs.de/order.in">
<title>IDOC.de Order Form</title>
<Order-No>1234567</Order-No>
<Message-Date>19991128</Message-Date>
<Buyer-EAN>12345000</Buyer-EAN>
<Order-Line Reference-No="0121314">
<Quantity>250</Quantity>
</Order-Line>
<input type="checkbox" name="partial" value="allowed"/>
<text>Tick here if a delayed/partial supply of order is acceptable
</text>
<input type="checkbox" name="confirmation" value="requested"/>
<text>Tick here if Confirmation of Acceptance of Order is to be returned
by e-mail
</text>
<input type="checkbox" name="DeliveryNote" value="required"/>
<text>Tick here if e-mail Delivery Note is required to confirm details of
delivery
</text>
</Book-Order>
```

Program 16:   XML Sales Order data

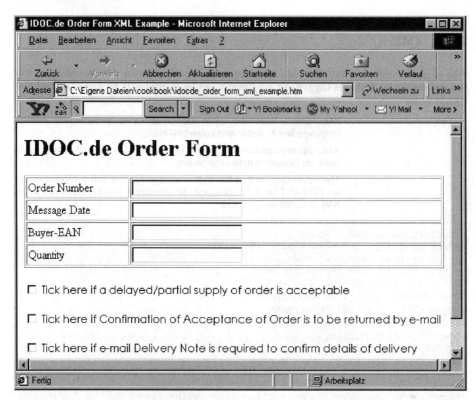

Illustration 45:   **XML Order form as displayed in a browser after interpretation by a JAVA applet**

XML plug-ins exist often as JAVA applets for standard browsers   The example shows some XML sales order. In order to be displayed with a standard browser like Internet Explorer 5, there exist plug-ins and JAVA applets that interpret the XML and translate the XML specific data tags into HTML form.

15.4 ANSI X.12

This is an example of how an ANSI X.12 EDI message for a sales order looks like. The examples do not show the control record (the "*envelope*"). EDIFACT looks very much the same.

The example describes a sales order from customer 0111213 for 250 KGM. The fields of a segment are separated by an asterisk (*).

We start with a header record describing the type of message (850). IDocs would store this information in the control record.

ST*850*000000101~	
ST01	Transaction 850 = Purchase Order
ST02	Set control number 453

Signal begin of transaction and identifies sender

BEG*00*NE*123456789**991125**AC~	
BEG01	00 - Original transaction, not a resend
BEG02	NE - New Order
BEG03	PO Number 123456789
BEG04	VOID
BEG05	PO Date 25/NOV/1999
BEG07	Client requests an acknowledgment with details and changes

Bill-to party and Ship-to party

N1*BT***0111213~	
N101	Bill to (VBPA-PARVW)
N104	0111213 number of bill-to-party (VBPA-PARNR)

N1*ST***5566789~	
N101	Ship to (VBPA-PARVW)
N104	5566789 (VBPA-PARNR)

The item segments for item 01 – 250 kg of material MY1001 for $15.3 per kg

PO1*1*250*KGM*15.3*SR*EAN*MY1001~	
PO101	Line item 1 – VBAP-POSNR
PO102	Quantity 250 - VBAP-KWMENG
PO103	Units Kilogram VBAP-MEINS
PO104	$15.30 - VBAP-PREIS
PO106	EAN – Material number
PO107	MY1001 (VBAP-MATNR)

Summary information to verify completeness

CTT*1*2~	
CTT01	1 PO1 segments
CTT02	2 some of quantities (ignore unit)

SE*7*000000101~	
SE01	7 segments altogether
SE02	Control number 453. This is the same as ST02

16

EDI Converter

R/3 does not provide any tool to convert IDocs into international EDI format like ANSI X.12, EDIFACT or XML. This conversion needs to be done by an external add-on product which is provided by a variety of companies who specialized in general EDI and e-commerce solutions.

Summary

- R/3 does not provide conversion to EDI standard formats like X.12, EDIFACT or XML

- Converters exist on UNIX and PC platforms

- Many converters are simple PC programs

- R/3 certification does only guarantee that the converter complies to RFC technology and works fine with standard IDoc scenarios

- Real life situations require a flexible and easily adaptable converter program

16.1 Converter

SAP R/3 has foregone to implement routines to convert IDocs into international EDI standard formats and forwards those requests to the numerous third party companies who specialize in commercial EDI and e-commerce solutions..

Numerous EDI standards

Nearly every standard organization defined an own EDI standard for their members. So there is X.12 by ANSI for the US, EDIFACT/UN adopted by the United Nations Organization UNO or XML as proposed by the internet research gurus of W3C.

Big companies define their own standards or dialects

But there is still more about it. Every major industry company defines an additional file format standard for their EDI partners. Even if they adhere officially to one of the big standards, they yet issue interpretation guidelines with own modifications according to their needs.

If a company does not play in the premier league of industry or banking companies, it will have to comply with the demands of the large corporations.

A converter needs to be open and flexible

As this leads to the insight, that there are as many different EDI formats as companies, it is evident that an EDI converter needs to have at least one major feature, which is *flexibility* in the sense of openness towards modification of the conversion rules.

There are hundreds of converter solutions on the market not counting the individual in-house programming solutions done by many companies.

EDI is a market on its own. There are numerous companies who specialized in providing EDI solutions and services. The majority of those companies do also provide converters.

Many of the converters are certified by SAP to be used with R/3. However, this does not tell anything about the usability or suitability to task of the products.

16.2 A Converter from Germany

In the forest of EDI converters there is only a very limited number of companies who have actual experience with R/3. We have chosen one very popular product for demonstration here.

Certification does not guarantee usability

Many of the converters are certified by SAP to be used with R/3. However, this does not tell anything about the usability or suitability to task of the products. The R/3 certificate is not a recommendation by SAP, hence it is only a prove of compliance to technology requirements.

Flexibility

Many of the converters have major deficiencies. It is e.g. important that the conversion rules can easily be changed by the permanent service staff of the client.

Graphical monitor

A graphical monitor that can handle both the converter and the R/3 is more than desirable.

Import IDoc definitions via RFC into the converter

In big EDI projects you also appreciate a tool that allows to import R/3 IDoc definitions into the converter. Using RFC the import should be possible without downloading a file from R/3.

Converter developed with R/3 in mind

The solution which made as smile is provided by the German company Seeburger GmbH http://ww.seeburger.de ..The company is different from most EDI service providers as it has its roots in R/3 consulting, so the folks have the viewpoint from R/3, while others see R/3 only as a data supplier.

EDI monitor

The product is made of several modules, among them you find a sophisticated EDI monitor to survey timely sending and reception of data.

Graphically map data structures

While monitors are common to most EDI converter solutions, our interest as developer focuses on the EDWIN editor. It allows to graphically map one data structure to a standard, to assign rules etc.

The illustration gives an idea of the editor. The tool can read the IDoc segment definitions from the R/3 repository via RFC and store back modifications if this should be necessary.

Easy to use

All in all we have chosen EDWIN as a stare-of-the art product for converter design with respect to versatility and ease-of-use.

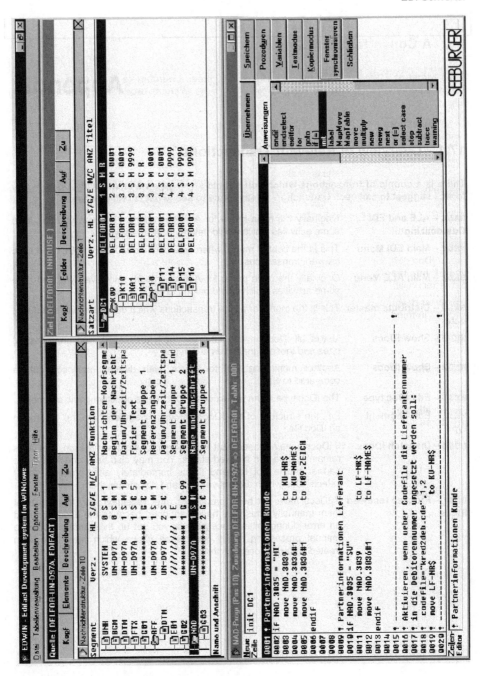

	17

Appendix

17.1 Overview of Relevant Transactions

There is a couple of transactions which you should know when working with IDocs in any form. I suggest to call each transaction at least once to see, what is really behind.

SALE – ALE and EDI Customizing
Originally the main menu for ALE operations. However, you find here some activities which would belong to WEDI as well.

WEDI – Main EDI Menu
This is the central menu, where you can find most of the EDI and IDoc relevant transactions.

BALE – Main ALE Menu
Originally the main menu for ALE operations. However, you find here some activities which would belong to WEDI as well.

BALM – Distribute master data
This is the menu for all the transactions which distribute master data.

WE05 – Show IDocs
List of all IDocs in the database, both processed, and unprocessed ones and those signalled as erroneous.

WE02 – Show IDocs
Another monitoring tool for IDocs, with different selection criteria compared to WE05.

WE30 – Edit IDoc type
The IDoc type is the syntax of an IDoc, i.e. is structuring into segments.

WE31 – Edit Segment type
Edit the structure of the IDoc segments. Segments are the records in an IDoc file.

BD88 – Dispatch IDocs
If IDocs have not been sent to a partner for whatever reason, the IDocs remain in a certain blocked status. You may also have checked in the partner profile not to send IDocs immediately after creation. BD88 selects IDoc which are not yet sent and dispatches them.

BD87 – Process received IDocs
If IDocs have not been processed after reception, you have to treat them manually. This may have happened because the IDoc signalled an error during initial processing or you set up the partner profile, to manual processing. BD87 selects all IDocs which have not been treated yet and processes them.

17.2   Useful Routines for IDoc Handling

These are some very useful routines, that can be used in IDoc processing.

Function IDOC_CTRL_INBOUND_CONVERT

Convert an IDoc control record into internal format

Convert an IDoc control record with structure EDIDD into the version dependent format EDI_DC or EDI DC40.

Function IDOC_DATA_INBOUND_CONVERT

Convert an IDoc control record into internal format

Convert an IDoc control record from the version dependent format EDI_DC or EDI_DC40 into the version independent format with structure EDIDD.

Function IDOC_INBOUND_FROM_FILE

Read a file and treat it as an IDoc

This function reads a specified file and handles it as an IDoc package. It stores the IDoc to the IDoc base and processes it according the preset customizing.

Function EDI_DATA_INCOMING

Read a file and treat it as an IDoc

Same as Function IDOC_INBOUND_FROM_FILE . This one has additional parameters, especially it allows using logical names instead of a physical filename

Read a file and treat it as an IDoc

This function reads a specified file and handles it as an IDoc package. It stores the IDoc to the IDoc base and processes it according the preset customizing.

Function IDOC_INBOUND_SINGLE

Central IDoc processing routine

This is an RFC capable function module, which takes an IDoc and its control record as a parameter, stores the IDoc to the IDoc base and processes it according the preset customizing.

Function IDOC_INBOUND_SYNCHRONOUS

Predecessor of IDOC_INBOUND_SINGLE for version 3.x.

Function OWN_LOGICAL_SYSTEM_GET

The routine reads the name of the logical system, on which the program is running. This is currently the entry found in table T000-LOGSYS. .

Function MASTERIDOC_DISTRIBUTE

Sends an IDoc immediately to the port according while making use of the appropriate customizing settings.

17.3 ALE Master Data Distribution

The ALE functionality comes with a set of transaction which allow the distribution of important master data between systems. The busiest argument for installing ALE might be the distribution of the classification from development to production and back.

R/3 comes with many predefined ALE scenarios for master data

ALE is only a mean to distribute IDocs in a controlled and event based manner. Here is a collection of the transaction which come already with R/3 and can be used to distribute data via ALE.

You can always create own ALE IDoc routines

If your master data is not with the standard functionality you can of course create your own function module to add on to the ALE mechanism.

MATMAS – material master

The easiest way to exchange material master data between systems or clients. The program is insensitive for the complex material views and screen sequence controls due to using the function `MATERIAL_MAINTAIN_DARK`.

DEBMAS - debtors

Debtor master data, tables `KNA1` etc.

CREMAS – creditors

Creditor master data, tables `LFA1` etc.

Classification

ALE is perfect to distribute the classification system to another system or client. The provided routines distribute nearly everything from the class definitions (tables `KLAH`) up to the characteristic assignment (`KSML`) and dependencies for variant configurator.

The dependency knowledge function modules you might have written are not distributed via ALE, because they are part of development.

The class 036 for dependency characteristics, the classification is refused from being distributed explicitly. Refer to http://idoc.de for a modification which allows you to send class 036.

17.4   Monitoring IDocs

There are some utilities in R/3 that help monitoring all the IDocs in the system. They allow viewing them, analysing eventual cause of error and retrying IDoc processing in case of failure.

WEDI - EDI central	The IDoc monitoring tools can all be accessed from menu *WEDI*.
WE02 – **WE05** **IDoc listings**	Transaction *WE05* and *WE02* display IDocs, which are found in the system; they allow to check if IDocs have been treated successfully or why they have failed.
BD87 - EDI central	*BD87* allows to process inbound IDocs again, if they have failed for some reason.
BD88 - EDI central	*BD88* allows dispatching outbound IDocs if they are stopped for some reason.
BD88 - EDI central	*WE19* is the IDoc test tool. It allows you to reprocess an IDoc in different manners for testing purposes. The test tool copies an existing IDoc and then calls the processing routine. Optionally you can call the processing in debug mode.

http://idocs.de http://logosworld.de

17.5   WWW Links

These is a random listing of interesting web sites dealing with the EDI topic. They are accurate as of November 1999.

http://idocs.de	The home page associated with this publication; updated program codes and FAQ related to EDI and SAP.
Data Interchange Standards Organisation	http://polaris.disa.org/; A page that reads about the multiple e-commerce standards with excellent links.
ANSI X12	http://www.x12.org/; Home page of ANSI X.12 standard with good glossaries and reference section
UN/EDIFACT	http://www.unece.org/trade/untdid/: the UN reference page on EDIFACT; just as chaotic as the whole standard
XML reference from W3C	http://www.w3.org/; the only reference to XML
XML/EDI	http://www.geocities.com/WallStreet/Floor/5815/ ; a good site on XML for use with EDI
More on XML and e-commerce	http://www.commerce.net/; deals with EDI for e-commerce
BISAC X12 EDI Cookbook	http://lbbc.lb.com/bisac/ ; gives you an idea of what X.12 is
IMA Links page	http://mlarchive.ima.com/ ; many links to related issues and a discussion forum

Ich interessiere mich für weitere Themen im Bereich Computing:

❑ Business Computing
❑ Informatik
❑ Wirtschaftsinformatik
❑ Mathematik

Bitte schicken sie mir kostenlos ein Probeheft:

❑ DuD
  Datenschutz und Datensicherheit
❑ Wirtschaftsinformatik

Ich bin:

❑ Dozent/in
❑ Student/in
❑ Praktiker/in

Bitte in Druckschrift ausfüllen. Danke!

Hochschule/Schule/Firma

Institut/Lehrstuhl/Abteilung

Vorname

Name/Titel

Straße/Nr.

PLZ/Ort

Telefon*

Fax*

Geburtsjahr*

Branche*

Funktion im Unternehmen*

Anzahl der Mitarbeiter *

Mein Spezialgebiet*

* Diese Angaben sind freiwillig.

Wir speichern Ihre Adresse, Ihre Interessen-
gebiete unter Beachtung des Datenschutz-
gesetzes.

322 01 200

vieweg

17.6   Questionnaire for Starting an IDoc Project

This is a sample questionnaire with important questions that need to be cleared before any development can be started.

To let me better estimate the actual amount of work to be done please answer the following questing carefully		http://logosworld.de http://idocs.de

If you do not know the answer, say so; no guesses, please, unless explicitly marked as such.
Development can only be efficient if the subsequent questions can be answered.

SAP R/3 release   ▢

Direction of EDI Solution	
❏	Inbound
❏	Outbound
Describe the partner system	
❏	R/3 Release: _____
❏	R/2 Release: _____
	if others
❏	Is data sent in SAP Idoc format?
❏	Is data sent in EDIFACT/XML/X.12 etc.? (-> we will need an EDI converter)
❏	Is data sent as a structured file?
Can standard Idocs be used?	
❏	Yes
❏	No
❏	Do not know
If Inbound: Can data be Inbound manually via a transaction only using provided Idoc data?	
❏	Yes
❏	No (-> then you have a customising problem to be solved beforehand)
❏	Do not know (-> try it out!)
If Outbound: can you see all the data to be sent somewhere on an SAP screen?	
❏	Yes
❏	No
❏	Do not know (-> try it out! Can only sent, what is displayed)
How many different Idocs will be sent (eg. No of files with different structure) ▢	

SAP Application area involved		*If you transactions involved, please list*
❏	SD customer orders create	❏ VA01 ❏ other: _____
❏	SD customer orders change	❏ VA02 ❏ other: _____
❏	SD delivery create/change	❏ VL01/VL02 ❏ other: _____
❏	SD picking confirmation	❏ VL01/VL02 ❏ other: _____
❏	Purchase orders send	❏ ME21 ❏ other: _____
❏	Customer Master	❏ VD01 ❏ other: _____
❏	Creditor Master	❏ KD01 ❏ other: _____
❏	Product catalogue	❏ MM01 ❏ other: _____
❏	Others, please describe	❏ other: _____

If Inbound: Do you have sample Idoc data already in a file	
❏	Yes
❏	No (go and get them! The first thing we would need)
❏	Do not know (-> sorry?!?, are you serious?)

Index

http://idocs.de http://logosworld.de

http://idocs.de http://logosworld.de